Production, Processing and Marketing of Milk and Milk Products

ABOUT THE AUTHOR

Dr. Rajendra Kumar Pandey

B.Sc. (Ag.) Gold Medalist, M.Sc.(Ag.) Gold Medalist and Ph.D. Animal Husbandry & Dairy Science, Ph.D. Econmics, BHU, DCH, ICRE, The Netherlands.

Former Head, Department of Animal Husbandry & Dairying. In-Charge, BHU Dairy Farm, Founder Coordinator, IGNOU, Study Centre No. 27109 in the BHU, Course Director ICAR Winter and Summer Schools. Presently working as a professor in the Department of Animal Husbandry & Dairying, Banaras Hindu University

Dr. Pandey has about 34 yers of experience as a research worker and teacher in the field of Animal/Livestock Production & Management. Dr. Pandey has undergone specialized training in Dairy Cattle Husbandry and Rural Extension Courses at International Agriculture Centre (IAC). Wageningen, The Netherlands.

Dr. Pandey is a Life Member of thirteen Academic Associations at National and International level. He is also Executive Member and Editorial Board Member of many Research Journals.

Production, Processing and Marketing of Milk and Milk Products

Editor
Dr. Rajendra Kumar Pandey
Professor & Former Head
Department of Animal Husbandry & Dairying,
Institute of Agricultural Sciences,
Banaras Hindu University,
Varanasi – 221 005, U.P.
India

BIOTECH
2025
BIOTECH BOOKS®

ISBN 978-81-7622-263-1

Published by: **BIOTECH BOOKS®**
4762-63/23, Ansari Road, Darya Ganj,
New Delhi - 110 002
Phone: +91-011- 43003222
E-mail: biotechbooks@yahoo.co.in

PRINTED IN INDIA

Preface

India has achieved a notable success in milk production with enhanced processing capability and marketing skills. Many of our dairy products have registered their presence in global market. The spectacular growth of dairy industry has been possible through sustained efforts of our scientist, planners, professionals and farmers and by creation of infra-structure for milk procurement and processing in various sectors, Many distinct brands have been established for pasteurized milk, sterilized milk, powders, ghee, infant foods, cheese, dairy whiteners, malted foods, butter and ice-cream. Casein and whey protein products are also produced in India. These products are generally manufactured using classical industrial processes developed elsewhere. Research programme in past were pursued to standardize the technologies for buffalo milk processing for the above named products. The Organized Dairy sectors in India, however, handles only 15 per cent of the total milk produced and much needs to be obtained in this regards.

Although, the traditional Indian dairy products enjoyed a great mass appeal because of their familiarities, understandings of the associated nutritional value and availability of good range of product categories at affordable prices, ironically limited processing technologies have been developed for commercial production of highly popular products like khoa, chhana, paneer, chakka and sweets derived there from. It may be of interest to mentioned that wide range of region specific traditional milk sweets of the different varieties, taste, texture and huge utilized nearly 50 to 55 per cent of milk produced in India. The small scale techniques used in production lead to wide variations in chemical and microbiological qualities. Such methods needs to be upgraded and modernized enabling the dairy industry to adopt them for industrial application to match increasing demand in domestic and Global markets.

Many useful machines have been developed for batch and semi-continuous production of khoa, paneer, chhana, ghee and srikhand with some efforts for packaging and self-life extension. Such works need to be intensified. The facilities should be adequate and capable to convey technology, techniques, knowledge and information to the smallest and poorest. Otherwise the efforts made and results achieved by research and development would remain consigned to the laboratories. Integrated organization in basic to bringing dairy industry out as a vibrant and commercial structure that can benefit the poor, raise their status and reduce poverty.

This book is a sincere effort to extend the recent knowledge and developments in the field of dairy farming and industry as a whole for milk production, processing and marketing strategies and is increasingly emphasizing on future growth of the dairy training facilities itself with the national dairy sector.

I am sure that the people engaged in this sector at any capacity, such as, scientist, researchers, students and ultimately farmers are the target group, would take stock of the innovations and identify the gaps for further planning future teaching and research objectives for rural development in eastern region through dairy production processing and marketing network in the form of the co-operative pattern and hopefully we can say the *"Second White Revolution through Eastern and North Eastern Region"*.

Dr. Rajendra Kumar Pandey

Contents

Chapter 1

Integrated Approach to Sustainable Dairy Development with Special Reference to North Eastern Region

S.S. Khare

Pro-Vice-Chancellor, North Eastern Hill University, Tura Campus, Tura

The cow is the foster mother of the human race. From the day of the ancient Hindu to this time, this kindly and beneficial creature has been one of the chief sustaining forces of human life. Milk and milk products were, are and will always be important sources of basic food nutrients for human being, be it child, young or old.

Traditionally cattle have been raised by people for prestige, as a subsidiary occupation to crop production in rural areas and for milk, fuel, cow dung and agricultural operations. However, through systematic planning by the Government of India in the form of various dairy development programmes starting from Key Village Scheme during the first five year plan, the Intensive Cattle Development Project introduced in sixties, operation Flood-I in 1970, Operation Flood-II in 1982 and Operation Flood-III in nineties, India has reached the present milk production of around 92 million tones/ year topping the world in terms of milk production as against meagre 17 million tones in 1950. The credit for such enhanced milk production goes to mainly cooperative movement apart from artificial insemination, better feed and fodder, processing, packing and marketing. During the Operation Flood-I, one million farm families were benefited through 9000 village level dairy cooperative societies in 18 milk sheds in 10 states. In each milk shed an infrastructure has been created to collect and process the milk and provide technical inputs for rapid increase in milk production. As it stands today more than 1 lakh dairy cooperative societies are working having more than 11 million farm members with 170 milk shed procuring about 18 million litre of milk per day. These 170 milk sheds are spread over 250 districts in 23 States and Union Territories.

Anand pattern cooperative system has proved to be the most suitable integrated system for dairy development, which consists of four tiers Primary Village Level Societies of milk produces constitute the first tier, key functional units consisting of district level cluster unions of such village level societies constitute the second tier and the apex state level federation of these cluster unions is the third tier. Lately different state-level federations are being linked together under fourth tier through the National Body namely National Cooperative Dairy Federation of India that will intermediate between State-level federations and oversee in its totality the development of dairy economy and movement of milk and milk products on country wide basis. The structure represents a compact and vertically integrated chain encompassing all operations intermediate between the primary producers and final consumers embracing the operations of milk production, collection, processing, packing and marketing. The bottom tier is mainly concerned with production, the middle tier is concerned with procurement and processing and the apex tier is concerned with marketing, balancing and quality product development and aims at giving better benefits to the milk producers. The profit gained goes to not only milk producers but also goes in improving quality of life of villagers.

The secret of success of this type of cooperative movement is that the ownership of this entire structure is in the hands of primary milk producers. While policy direction is in the hands of democratically elected representatives in each tier, technical management at second and higher level is provided by paid professionals, managers who happen to be employees of the cooperative and are accountable to the respective elected boards.

On one side it is a matter of pride for each of us that India is top most in milk production but on the other side it is a matter of concern that the milk production per cattle per day is hardly 3 kg/day as against 25 kg/day for average cattle in U.S.A. This is because of about 80 per cent of cattle in India being non-descript give hardly 1-2 kg/day. Further, per capita availability of milk every day at present is about 230 gm against 8.5 kg/day in New Zealand. This shows that there is a tremendous scope of enhancing milk production by encouraging farmers to raise high breed cattle like Jersey, Holstein-Friesian, Brown Swiss, or ethnic cattle breeds like Sahiwal, Sindhi, Gir, Deoni, Rathi, Tharparkar etc. In fact, as compared to Jersey, Holstein-Friesian, Brown Swiss, cattle of the breed Sahiwal, Sindhi, Gir, Deoni, Rathi, Tharparkar etc. may be giving some what less milk as compared to Jersey, Holstein-Friestan etc. but they are more sturdy, demanding less day to day care and maintenance and provide sturdy bullocks for agricultural operations also. The scenario of dairy development in N.E Region is all the more discouraging and is much worst in Meghalaya.

The average per capita availability of milk per day in N.E. Region is meagre 90 gm as against 230 gm for the country. The only cooperative society some what in line with AMUL is KOMUL in Nagaland and WAMCUL in Assam. The total milk production of all the 8 states of N.E. Region is hardly 1.5 million tones/year which comprises only 1.6 per cent of the total milk production of the country and total cattle in N.E. Region is about 7 per cent . This shows that the average milk yield of cattle in N.E. Region is much below the average milk yield of the cattle in India. There are very few dairy plants registered under MMPO in N.E. Region and very few milk processing units. The farmer members of the dairy societies are only 0.2 per cent of the total farmer members of dairy cooperative societies in the country. I am of the opinion that the main reasons of such bad scenario of milk production are (i) lack of cooperative movement similar to that of Anand (ii) lack of awareness about the rich potential dairy industry has for employment and economic upliftment of the region (iii) lack of awareness and availability of high milk yielding breeds of cattle (iv) lack of awareness and availability of quality feed and fodder (v) lack of milk collection centres (vi) lack of chilling units and processing units (vii) lack of awareness and availability of artificial insemination (viii) lack of awareness about health care of cattle and lack of veterinary clinic and veterinary health centres, (ix) lack of marketing net work (x) lack of communication facilities and (xi) lack of skilled manpower.

It will not be out of way to mention that the total annual consumption of milk and milk products of Amul Taja and of some companies of West Bengal in N.E. Region is more than worth Rs 60 crores. Because of lack of availability of fresh milk and its products, people here are compelled to go for Amul Taja milk at the rate of Rs 27.00 per litres. If we can increase milk production in N.E. Region, people will certainly prefer to consume fresh milk and its products at much cheaper rate than Amul Taja and that of other brands.

Table 1.1: Annual Milk Production in Different States of NE Region (000' tones)

Sl.No.	States	1981-82	1991-92
1.	Assam	469.00	639.0
2.	Manipur	62.00	83.0
3.	Meghalaya	58.00	50.0
4.	Nagaland	3.45	44.0
5.	Sikkim	18.00	29.0
6.	Tripura	17.50	32.0

Table 1.2: Estimates of Milk Production in NE Region 1997-98 (000' MT)

Sl.No.	States	Milk Production
1.	Arunachal Pradesh	44.0
2.	Assam	775.0
3.	Manipur	62.0
4.	Meghalaya	59.0
5.	Mizoram	20.0
6.	Nagaland	46.0
7.	Sikkim	35.0
8.	Tripura	47.0

Table 1.3: Projected Milk Production upto Year 1998-99 to 2005-06 (000' tones)

Sl.No.	*States*	*1998-99*	*2002-03*	*2004-05*	*2005-06*
1.	Assam	815.32	940.58	1010.25	1047.00
2.	Manipur	99.73	112.72	119.84	123.56
3.	Meghalaya	54.88	56.88	57.91	58.44
4.	Nagaland	105.85	190.04	254.64	294.75
5.	Sikkim	40.49	49.45	54.65	57.45
6.	Tripura	47.21	58.75	65.54	69.23

Table 1.4: Share of Cow, Buffalo and Goat Milk to Total Milk Production in NE States (000' MT) – 1993-94

Sl.No.	*States*	*Cow*	*Buffalo*	*Goat*	*Total*
1.	Arunachal Pradesh	21.0	-	-	21.0
2.	Assam	561.0	91.0	24.0	676.0
3.	Manipur	63.0	21.0	-	84.0
4.	Meghalaya	50.0	3.0	-	53.0
5.	Mizoram	9.0	-	-	9.0
6.	Nagaland	42.0	2.0	1.0	45.0
7.	Sikkim	NA	NA	NA	30.0
8.	Tripura	NA	NA	NA	35.0

Table 1.5: Per Capita Availability of Milk 1997-98 in NE States (gms/day)

Sl.No.	*States*	*1993-94*	*1997-98*
1.	Arunachal Pradesh	62.0	118.0
2.	Assam	78.0	82.0
3.	Manipur	118.0	79.0
4.	Meghalaya	77.0	78.0
5.	Mizoram	32.0	64.0
6.	Nagaland	93.0	84.0
7.	Sikkim	186.0	191.0
8.	Tripura	33.0	40.0

Table 1.6: Project per Capita Availability of Milk in NE States in 2004-05 and 2005-06 (gms)

Sl.No.	States	2004-05 (gms)	2005-06 (gms)
1.	Assam	93.53	94.88
2.	Manipur	135.37	136.62
3.	Meghalaya	67.72	66.88
4.	Nagaland	436.87	495.01
5.	Sikkim	279.01	287.01
6.	Tripura	49.33	51.00

Note: Requirement of milk has been worked out on the basis of recommended norms of 220 gms per day per capita by the ICMR.

Table 1.7: Average Milk Yield of Cows, Buffaloes and Goats in NE States in 1991 (in litre/day)

Sl.No.	States	Cow		Buffalo	Goat
		Non-descript	Cross-breed		
1.	Arunachal Pradesh	1.030	4.56	NA	NA
2.	Assam	0.912	3.74	1.770	0.227
3.	Manipur	1.400	7.96	3.16	NA
4.	Meghalaya	0.720	8.410	1.96	NA
5.	Mizoram	0.710	8.500	NA	NA
6.	Nagaland	1.710	6.760	3.160	0.380

Table 1.8: Total Number of Cattle, Buffalo and Goat in NE States in 1992 (in 000)

Sl.No.	States	Cattle	Buffalo	Goat
1.	Arunachal Pradesh	324.0	5	125
2.	Assam	10118.0	959	3454
3.	Manipur	719	114	38
4.	Meghalaya	535	33	195
5.	Mizoram	59	7	21
6.	Nagaland	332	32	151
7.	Sikkim	198	2	114
8.	Tripura	949	20	429

Table 1.9: Annual Growth Rate of Milking Cows and Buffaloes in NE States–1992

Sl.No.	*States*	*Annual Growth Rate of Milking Cows (per cent)*	*Annual Growth Rate of Milking Buffaloes (per cent)*
1.	Arunachal Pradesh	(-) 3.58	(-) 12.94
2.	Assam	6.75	1.031
3.	Manipur	4.56	0.00
4.	Meghalaya	0.00	0.00
5.	Mizoram	8.45	0.00
6.	Nagaland	8.76	14.87
7.	Sikkim	1.61	0.00
8.	Tripura	11.46	3.13

Table 1.10: Milk Production Status in NE States in 1998-99

Sl.No.	*States*	*Milk Production (000'MT)*
1.	Assam	44.5
2.	Assam	798.0
3.	Manipur	65.0
4.	Meghalaya	61.0
5.	Mizoram	21.5
6.	Nagaland	47.5
7.	Sikkim	40.0
8.	Tripura	45.0

Although due to various Government schemes and due to the cooperative movement of Anand pattern, there has been significant enhancement of milk and its product, but there have been some weaknesses also, perhaps due to improper implementation of some aspects. Cross bred female cattle are very good for milk production if they are given quality feed and fodder, required minerals and are given proper health care. But cross bred male cattle are not suitable for agricultural operations. One of the main fall out of indiscriminate artificial insemination in many pockets of the country has been conspicuous decrease of draught cattle, as cross-bred bullock were slaughtered, since found unfit for ploughing. Therefore, the policy makers have to consider maintaining disease resistant and sturdy bullocks suitable for agricultural operations along with producing exotic high milk yielding milch cattle. Another weakness has been in terms of failure of dairy cooperatives to evoke equal level of success in all the states and also presence of very few dairy cooperative societies and very few farm members in many states. Effort has to be rendered to have more and more effective dairy cooperatives with more and more farmer members in N.E. Region and other areas where there are none and to analyse the weaknesses of existing dairy cooperatives which are not giving expected results and take corrective measures.

Table 1.11: No. of Milk Processing Factories and No. of Liquid Milk Plants as on 31.3.1997 in NE States

Sl.No.	States	No. of Milk ProcessingFactories	No. of Liquid Milk Plant
1.	Arunachal Pradesh	Not reported	1
2.	Assam	9	9
3.	Manipur	-	1
4.	Meghalaya	-	6
5.	Mizoram	-	1
6.	Nagaland	1	1
7.	Sikkim	2	2
8.	Tripura	1	1

Table 1.12: State Dairy Cooperatives in NE States as on 31.3.1998

Sl.No.	States	Dairy Coop. Societies, Organized (No)	Farmer Members ('000)	Rural Milk Procurement (000 kg/ day)	Milk Marketing (000 l/day)	Processing Capacities (000 L/day)
1.	Assam	123	2	3	6	60
2.	Nagaland	35	1	1	1	-
3.	Sikkim	162	5	7	7	15
4.	Tripura	84	4	1	5	10

Table 1.13: No. of Dairy Plants Registered Under MMPO as on June 1, 1996 in NE States (capacity 000 lit/day)

States	*Cooperative*	*Private*	*Others*	*Total*
Mizoram	1	-	-	1
Tripura	1	-	-	1
Sikkim	13	-	-	13

Table 1.14: Capacity of MMPO Registration by Central Registering Authority as on March 1991

States	*District*	*Name*
Sikkim	East Sikkim	Sikkim Co-op. Milk Producers Union Ltd.
Tripura	East Tripura	The Tripura Dist. Milk Producers Union Ltd.

For sustainable dairy development following measures need to be taken.

1. Raising of buffaloes should be encouraged as buffaloes need much less care, more disease resistant and do not need very high quality feed and fodder for producing good amount of milk as compared to cows. With them the cost of milk production reduces.
2. Indigenous and non-descript cattle giving less milk should be replaced in phases by high milk yielding cattle by artificial insemination. But in this process a balance has to be kept to ensure rugged and strong bullocks fit for agriculture operations. For this biotechnology and embryo transfer technology for production of strong bulls may be used. Techniques are now available to produce more than one offspring from the same female through open nucleus breeding system. With the help of hormones, high milk yielding female cattle is made to produce multiple ova, which can be frozen for future use. The female cattle giving low milk yield are used as mothers for implantation of fertilized embryos from high milk yielding females. Combination of multiple ovulation, embryo transfer and semen from progeny tested bulls can give faster genetic gains. Along with this farmers may be encouraged to raise

indigenous cattle with some what high milk giving capacity, as the cost of milk production with such cattle is less.

3. The dairy farmers may be encouraged to grow berseem, lucerne, cowpeas and other crops and fodders required for enhancing milk capacity of cattle. Instead of growing these crops as main crops, the farmers may be advised to grow them as inter crops. In addition, farmers may be advised to grow crops like sugarcane, sunhemp, carrot, cauliflower, turnip which provide good quality by products and these by-products can be utilised as a source of feed to the cattle. Anjan, rhodes, stylo and siratro grasses can be grown in guava, citrus or mango orchards. Similarly in coconut or banana plantation, one can grow fodders like cowpea, sunhemp and pillipsera etc. It is surprising that less than 3 per cent of the cultivated land is being used for fodder. It needs to be increased to at least 8 per cent . In N.E. Region this percentage is hardly 1 per cent . In fact farmers having no land or less land may be encouraged to go for dairy instead of rain fed agricultural crop production which involves much more risk in terms of flood or drought and the income is also not on day to day basis. Technologies for improving agro by products as feeds are available and are worth using. Urea molasses treatment improves the quality of dry fodders and straws of wheat and rice. By using the proteins that bypass the rumen, the nutrients available to the animals can be increased and hence becomes more economical. There are many additives, hormones and micro-nutrients that can be used to improve the bio-availability of nutrients in the feeds and fodders given to animals.

4. Milk production with high breed animals can be increased by proper health care and management which can be done in three tier system. First tier is the first aid at village level through a local resident trained as Veterinary Assistant enable to diagnose and provide immediate relief to sick animals. If required, he/she should be able to contact regular veterinary clinic installed in every block. Second tier may be Mobile veterinary service which should visit the villages on predetermined days, time and place in the

village known to the cattle owners of the village. The mobile service should be able to treat most of the diseases like mastitis, helminth, infestations, micro-mineral deficiencies, parasitic infestations, tuberculosis, para-tuberculosis and poxes etc. along with follow up instructions to the local resident veterinary assistant. The third tier should be centre having facility of artificial insemination and regular vaccination against epidemics, infections and contagious diseases *e.g.* foot and mouth, rinderpest, haemorrhagic, septicaemia, brucellosis, anthrax, black quarter, acute undifferentiated diarrhoea of new born. Such centre may be established on block level by Government or NGOs or entrepreneurs. There are thousands of M.Sc. (Ag) Animal Science sitting unemployed. They can start such clinic/centre. There should be a concentrate feed mill, milk collection centre and a chilling centre in each block with the provision of tankers to carry chilled milk to milk processing centres installed in each district having proper arrangement of carrying packed milk and its product to distribution centres. This will help in increasing the present milk procurement of meagre 20 per cent to a significant level.

5. As most of the people in N.E. Region are regular meat eater, more emphasis may be given for production of fermented milk, yoghurt and other milk products.
6. Milk producers are scattered in small groups, mostly away from urban areas. Dairy cooperative societies similar to Anand pattern may be established in N.E. Region with four tiers enrolling more and more farmers as member of the village milk producers societies. The existing dairy societies should be strengthened by getting more farmers enrolled as members, by having better management and by providing better technical know-how at each level.
7. The communication and transport in N.E. Region and in many less developed pockets of India is very poor and needs improvement to facilitate better marketing.
8. The country is facing serious unemployment problem. Millions of educated youth including M.Sc. (Ag) are unemployed. Dairy industry has got a tremendous

potential to give employment to lakhs of educated youth. They can go for entrepreneurship on many aspects of dairy industry including small dairy unit, veterinary centre, Agro-clinic, milk procurement centres, milk chilling unit, concentrate production units, milk processing unit, marketing and transportation of milk and milk products. The educated youth having land can even go for production of high quality fodder. This will not only encourage self employment but also boost dairy industry. Many activities of dairy industry are suitable to housewives after their necessary household works.

9. The cattle and other animals of the farmers need to be insured against causalities.
10. Application of information, technology and computers has helped the leading cooperative in India to change the rural face. More than 3000 cooperatives in Gujrat and other states have stand-alone computerised automatic milk collection units. They capture information of fat content of milk, its volume, calculate the amount to make the payment on the spot. This has increased trust, transparency and efficiency of milk collection. Introduction of internet is going to link all e-transactions between the village cooperatives and processing units. Recently Geographical Information System (GIS) software is being used by Gujrat Cooperative Milk Marketing Federation at the village level for data on farmer members, their land holding and animals. Wireless communication can help in tracking milk carriers for break down, can net work the Veterinarians to quickly reach animals suffering in far flung rural areas. Thus proper telecommunication, internet, cybernetics and judicious use of information and technology can revolutionise dairy industries.

This is the age of globalisation and liberalization creating a single world market. With cheap labour and cheap raw materials, dairy industry in India has a potential to play a vital role in global market. But in practice, rich developed countries tend to dominate the underdeveloped and developing countries through their dictated policies. Most of the developed countries protect their dairy industries by providing high subsidies to their milk producers and by preventing import. They also provide high subsidies for export of milk products. This has created unhealthy and unfair competition in the world market and even in the domestic market of importing countries, India being one of them. The developed countries mount intense lobbying pressure at WTO meets to retain such protection and such subsidies. It is painful to mention that India has committed zero per cent base and bound rates on imports of skimmed and full cream milk powder and 40 per cent on butterfat, cheese and whey under the WTO agreement. No other country other than Singapore has agreed to zero per cent duty. Because of such unfair policies at the WTO level, the poor countries become poorer and rich become richer. This is despite cheap production of milk and milk products in our country. It is high time our policy makers in the country should take necessary measures to get rid off from this unfair deal.

Asian dairy products market is growing very fast. The total demand of milk and milk products in Asia is likely to increase from approximately 200 million metric tones at present to around 600 million metric tones in next 20 years. There is a great opportunity for milk growers in India to capture this market. The concept of food parlours is opening new vistas for ready to serve dairy products which would ride piggyback on the fast food revolution sweeping across urban India. Further, there is a growing market for dairy products as ingredients used as raw material on pharmaceutical and allied industries.

Coming to the scenario at world level, the most critical issue now confronting the world dairy industry is the fact that world milk production is not expected to be able to keep up with the increasing demand for dairy products. The farm-gate price of milk in India is one of the lowest in the world. There is a need of a strong cooperation amongst the developing countries, specially in Asia, to stand together at WTO to build pressure to create a fair level playing field in the world dairy market and to give equal opportunity to all the players for international trade and marketing. With the reduction of subsidies by developed countries like U.S.A., European Union and Canada etc., the situation has changed during the new millennium. India is a new entrant to the international dairy market and has a potential to be a significant player.

Branded western and ethnic dairy products are witnessing rising demand and increased acceptance, specially in urban areas.

The success of branded dahi, yoghurt, flavoured milk and traditional sweets such as shrikand, gulabjamun, peda and burfi amongst urban customers suggest potential for entering into Asian and international markets with such products. But while entering into international market, quality assurance of internationally accepted standard in production has to be kept.

Three phases of Operation Flood has certainly brought white revolution, but many pockets in the country including N.E. Region could not enjoy the impact of the white revolution. I take this opportunity to appeal dairy scientists, policy markers, Indian Dairy Board, NDDB, ICAR, KVK, NIRD, SIRD, Agricultural Universities, NGOs and other concerned organisations to jointly work for the fourth phase of Operation Flood with special emphasis on N.E. Region and other such regions which could not have the required impact of earlier phases of Operation Flood. The success of Operation Flood certainly lies in development of dairy industry and in economic development of our country through it, but the real success lies in making sure that each child gets at least a glass of milk every day and a good part of profit passes on to each milk producer particularly in rural areas, improving his quality of life. Being an optimistic person, I am sure that if all the concerned organisations work out jointly, there is no reason why dairy industry can not play a key role in placing our country amongst developed country by 2020.

With these words I inaugurate this Summer School and hope that the participants and speakers in the Summer School will ponder on various problems faced by dairy industry specially in N.E. Region and in some other less developed pockets of the country and try to come forward with solutions to these problems.

I wish the Summer School a grand success.

Dairy Cooperatives in North East States

Assam

1. Cachar and Karimganj Dist. Milk Producers Cooperative Union Ltd., Silchar.
2. DSD Cooperative Milk Producers Union Ltd., Jorhat.
3. West Assam Milk Producers Cooperative Union Ltd., Khanapara, Panjabari, Guwahati-781002.

Mizoram

1. Mizoram Coop. Milk Producers Union Ltd., C/O Directorate of Animal Husbandry and Veterinary, Govt. of Mizoram, Aizawl-796001.
2. Mizoram Multi-commodity Producers Cooperative Union Ltd., Aizawl, Mizoram-796017. Brand Name : MULCO Milk.

Nagaland

1. Bhola MPCS Ltd., Dimapur-797112 (Dairy Products and sale of milk).
2. Chumukedima III MPCS, Dimapur. (Buy and sale of milk).
3. Dairy Farms MPCS, Dimapur. (Sale of milk)
4. Everst MPCS, Dimapur (Buy and sale of milk)
5. Kohima Milk Union Ltd., Dimapur (Sale of milk)
6. Medziphema MPCS, Medziphema (Milk products and sale of milk)
7. NC MPCS, Kedima, Dimapur (Milk sale).
8. Tulovi Milk Producers Cooperative Society, Dimapur (Buy and sale of milk)
9. Valley Milk Producers Cooperative Society, Dimapur (Buy and sale of milk).

References

1. Mudgal, V.D., Singhal, K.K. and Sharma, D.D. Advances of Dairy Animal Production, International Book Distributing Co., 2003.
2. Ramasamy, D., Shibu, A.V. and Gopi, H., Dairy Technologist's Hand Book, International Book Distributing Co., 1999.
3. Venkatasubramanian, V., Singh, A.K., Rao, S.V.N. Dairy Development in India, Concept Publishing Co., New Delhi, 2003.
4. Winton, A.L. and Winton, K.B. Milk and Milk Products, Agrobios, India, 2000.
5. Meghalaya Socio-Economic Review : Government of Meghalaya, Directorate of Economics and Statistics, Shillong, 2003.

6. Where do we stand in 2003 Meghalaya and North East and India and the World : Govt. of Meghalaya, Directorate of Economics and Statistics, Meghalaya, Shillong, 2004.

7. Meghalaya Achievements in 30 years 1972-73 to 2001-2002; Govt. of Meghalaya, Directorate of Economics and Statistics, Shillong, 2002.

8. Pocket Statistical Hand Book Meghalaya, Directorate of Economics and Statistics, Shillong, 2003.

9. Statistical Hand Book of Manipur, Directorate of Economics and Statistics, Govt. of Manipur, Imphal, 2002.

10. Mizoram Statistical Hand Book 2000, Directorate of Economics and Statistics, Govt. of Mizoram, 2001.

11. Statistical Hand Book of Nagaland, 2000, Directorate of Economics and Statistics, Govt. of Nagaland, Kohima, 2001.

12. All India Dairy Business Directory (Dairy Year Book) 2001, Special Millennium Issue, Sadana Publishers and Distributors, 2nd Edn.

13. Encyclopaedia of Dairy Sciences, Edited by Hubert Roginski, John W. Fuquay and Patric F. Fox, A.P., 2003.

14. Indian Dairyman, Vol. XLI, No. 11, Nov. 1989.

15. Indian Dairyman, Vol. 56, No. 7, July 2004.

16. Indian Dairyman, Vol. 56, No. 10, Oct. 2004.

17. Indian Dairyman, Vol. 56, No. 12, Dec. 2004.

18. Indian Dairyman, Vol. 57, No. 2, Feb. 2005.

Chapter 2

Role of Marketing in Economic Development

B.D. Shukla

Retd. Professor of Agricultural Economics, B.H.U.

Recent changes in the macro-economic policy environment in the country has undergone a subset entail change. These change specially those relating to the liberalization of trade and Industry and new initiatives to link Indian economy with the global market necessitate a change in the role and perspective of the agricultural marketing system. Investment in agricultural sector, both in public and private sections, has risen. Agricultural production, in general has achieved reasonable growth rate. But the growth rate has not only to be maintained, but accelerated and fluctuations in agricultural production are to be minimized. The efforts are already under way to evolve location and product specific technologies transfer them to producers and assure input supply to the entrepreneurs in right time, place and quality. The rate at which new technology, and yield increasing inputs are adopted by the producers is affected by the prices of inputs and outputs, simultaneously, consumers also expect the availability of good lat reasonable prices for achieving these conflicting objectives, marketing system for agricultural commodities and inputs has to play a very crucial role. This necessitate an understanding of the marketing system and market structure for agricultural products and inputs by all the sections of the population, *viz.* producers, traders, consumers, extension workers, scientists, sociologists, administrators, planners and politicians. Therefore the performance of the existing marketing system, institutions and policy in accelerating agricultural development of the country is of great significance. Agricultural marketing plays an important role not only in stimulating production and consumption, but in accelerating the pace of economic development. Its dynamic functions are of primary importance in promoting economic development. As such, it is the most important multiplier of agricultural development. The technological break through in the form of H.Y.V. seeds, increasing use of fertilizers, insecticides, pesticides, pumping sets, tractorization and high yielding milk cattle has led to a substantial increase in production and thereby in larger marketable surplus. To maintain this tempo and PACE of increased promotion through technological break through, an assurance of remunerative prices to the producers and fair prices to the consumers is a pre-requisite and this can be activated through an efficient system of marketing.

The Agricultural marketing system plays a dual role in economic development in countries whose resources are primarily agricultural. It is the marketing system that transmits the crucial price signals. An efficient agricultural marking system leads to the optimization of resource use and output-input management. Efficient marketing can also contribute to an increase in the marketable surplus by sealing down the losses arising out of inefficient processing, storage and transformation consequences:

- ✰ *Increase in Producers Income*: By reducing middlemen, commission, market charges and containing malpractices. If the producer does not have an early accessible market outlet; he has little incentive to produce more.
- ✰ *Widening of markets*: Through efficient storage and transportation to the area of high demand from low demand, area ensuring better prices.
- ✰ *Growth of Agro base industries*: By supplying raw goods.
- ✰ *Price-signals*: for the products: where, when in what form to sell.

Adoption of New Technology

- ✰ *Employment*: To millions of persons in transportation, processing, packaging and storage.

Addition to National Income

- *Better living*: Quality and pure goods at fair prices.
- *Creation of utility*: Marketing is a part or advance stage of production. Production is complete only when the product reaches a place in the form and at the time required by the consumers. Marketing adds cost to the products; but at the same time, it add utilities to the product, which are of four types.
    1. *Place utility*: is created by shifting the produce from low demand to high demand area, hence fetches remunerative price to the producers. Thus place utility is a function of transportation: hence good link roads, vans and vehicles are needed for quick transport.
    2. The processing function adds form utility to the product by changing the raw material into a finished form. Thus the product becomes more useful and fetches high prices. *e.g.* oil from oil suds, rice, flour, maida, suji, butter, ghee, cheese, khoa, curd, sugar, cloth from cotton. The processed forms are more consumable than the original raw materials. Tremendous employment opportunities and saves the produce from being wasted.
    3. *Time Utility*: old rice, vegetables or products from main season in the lean/off season. Storage function add time utility. Cold-storage, warehouses save the product and make them available through out the year round, off-season products make life interesting and cheerful and provide better living standards.
    4. *Possession Utility*: The marketing function of buying and selling helps in the transfer of owner ship from one person to the another, products are transferred through marketing to persons having a higher utility from persons having low utility.
- *Growth of marketing*: No attention in I three 5 yr. plans. To begin with self interest and traditional agriculture did not warrant growth of marketing in Agriculture but due to National Commission on Agriculture land and due to.

1. *Specialization*: efficiency and out put increased.
2. *Urbanization*: 62.9 m. in 1951 to around 250 million in 2001, 217.2m. in 1991, 181.9 m. in 1981.
3. Transportation and communication.
4. Technological and administrative change in agriculture.

Chapter 3

A Profile of Dairy Industry and its Sustainability

Rajendra K. Pandey
Head, Department of Animal Husbandry and Dairying, Banaras Hindu University, Varanasi – 221 005

In order to understand what sustainable development means one need to make a distinction between growth and development. Growth is a quantitative increase coming mainly from the assimilation of products or of the incomes, while development is a qualitative change in the life of human being visible in their use of material goods, in their standard of living, in their education and in their prosperity our conventional method of estimating the growth is through GDP- that is Gross Domestic Products or through Gross National Products (GNP)

Every country produced a number of commodities and service. Each one of these can be assigned a value, and when we assimilate all these values, it becomes the gross domestic products. In the conventional method of estimating GDP, there is no concern for the improvement of life that has taken place. All these conventional method have a drawback that they do not take into account the loss of natural resources in the production process, either renewable or non-renewable, is utilized. But, in the conventional method, there is no account taken of the resources which are lost in the production process of these commodities. So we may be having a higher growth rate, but in fact it may not the adding to any development. Does the economic development lead to better human beings? Does the economic growth lead to better life on this planet for human beings. The answer is 'No'. We know of the contradictions, the great disparities that the economic growth has brought in this world. Today billions of people are living below the poverty line. Thousands of young children die daily due to malnutrition and diseases. One third of the world's adult population lives in India. Half of them are illiterate, two-third of them being women. Income wise also there are great disparities in the world. The rich of the world, per cent of what is poorest one-fifth of the world could take out. The industrialized nation consumes 10 times more energy than the developing nations. Also 70 per cent of the total waste and the emissions from the harmful oxides and other gases are discharged by the developed nations. So, has the economic growth really brought about any improvement in the living standards of the nations, particularly the poor ones.

In such a contradiction situation, what is the meaning of sustainable development? Why should we talk about sustainable development at all? The answer is because we have come to realize that economic growth itself is not going to solve the problems of the world. It is not going to attack poverty prevailing in the world. It is only through sustainable development that we can tackle the problem of poverty. The duty imposed by sustainability on us is that we should replace the natural endowment of the world. The natural endowments of the world are non-renewable resources, like various types of plants, animals, livestock and other species. In our attempt to achieve higher and higher economic growth we should not forget that these natural endowments are the means which will provide a better living to the coming generations. It is our duty that each generation leaves behind, each generation replaces what it takes away from nature so that the future generation can have a similar or a better living standards.

Indian Dairy Industry and Sustainability

In the light of these criteria, let us examine the scenario of the dairy industry in India. Indian Agriculture is an economic symbiosis of crop and cattle production. Small and marginal farmers own almost 80 per cent of the total land holdings. About 67 per cent of the workfare of 280 million is engaged in agriculture either as cultivator or farm labour. In a normal year, crop production can generate employment for this work force for only 90 to 120 days. For the remaining period, they are virtually unemployed. In this milieu, dairying sets right these imbalances in employment. The dairy sector today provides some 70 million farm families the triple benefits of nutritive food, supplementary income and productive employment for family labour, mainly women, clearly, any effective programme for poverty alleviation would be strengthened by dairying.

Thus the first unique characteristic of Indian Dairy Industry is that it is in the hands of small holders. Two third of the milk producers are small or marginal farmers, 20 per cent of the milk producers have no land, 66 per cent of them have land, but they are small or marginal farmers with land holdings of a couple of hectares. For fodder requirements of their cattle, the Indian farmers mainly depend on the agricultural waste and agricultural by-products, or on grazing. They do not utilize grains, thus the cattle do not compete with the human beings. This is one of the distinct features of the Indian dairy industry and different from the industry related world where dairying in primarily based on the utilization of the import which compete with the requirements of human being.

The second feature of the Indian Dairy Industry is that because milk production by small farmers they keep one or two milch animals only. Therefore, the collection of small quantities of milk from these farmers is a very cumber some process. This was the reason why these milk producers were organized into a cooperative society.

Cooperative milk industry has many challenges. Some are internal others are external. The intensity of these challenges is on the increase due to the open industrial policy. The main challenge is to face the private industry. Cooperatives have to go through the numerous hurdles of rules and regulations. To have to go through the rituals of following all these rules on the one side and to compete in business with the private sector which is immune to all the provisions of law is a Herculean task.

Moving Towards Sustainable Dairy Industry

The rural financial growths of this country during the last more than four decades have deep roots in the cooperative movement. In Gujarat, dairying prospered out of the farmers strong felt need to cooperatives. The milk procurement, processing products, and the marketing operations have a well-knit network. Against this an Indian milk producer has the orthodox thinking and vision since pre-independence day and he considered as animals are required for farming and manure is a good fertilizer. This resulted in, nominal milk production because there was no emphasis on improved breed of animals. Due to limited income from farming, the poor farmer had to deprive his children and family of the milk produced at home and sells it for making few more rupees. Marginal land-holding and lack of technical knowledge about the milk production were his other difficulties. However, his nominal income from milk proved to be a good support which has culminated in the cooperatives.

Cooperative Dairying

Collection of milk from all individual milk producers in the cooperative village level society and transporting it to the district level milk processing plant is a methodology adopted in the cooperative dairying. In the past or *even today also in the caster region, the milk production was fully dependent on the middlemen or an agent visiting the village for buying the milk.* If this did not happen for any reason milk being perishable, the only choice left to the producers was to consume in his household and convert the excess into butter and ghee by employing the traditional or crude methods those were known to him. These products were sold at throw away price. In contrast to this, today the milk producers is part of a national milk grid and of the latest technology and scientific way of milk production and processing.

Unfortunately, there is a general notion that milk is almost a free commodity, meaning the animals are free and the countryside grass is free. People conveniently forget that animal itself costs several thousand of rupees. Cost of rearing is not taken into account. Cost of feed and other inputs are overlooked.

Thus, sustainability is determined by the complex interaction of the biological physical and socio-economic factors that constitute the basis of all production systems. Sustainability should be treated as a dynamic concept making allowances for the changing need of steadily increasing global population. This requires a comprehensive approach to improve existing systems and to develop novel ones to avoid obsolescence. Expanding on this, it is not difficult to identify the four basic structures of dairy industry; in fact, these are common to most industries. The four structures are – production of milk, processing of milk, the consumer and human resources, what have been discussed above.

Thus, the Dairying is important for India as, it is rural based land saving and gender neutral. Compared to crop production, it has small favorable ratio of employment to output. Dairy farming makes a substantial source of income for the landless labour as well as the marginal farmers. It is scaled neutral and dominated by small herd sizes of two to four animals.

Cattle and buffaloes are integral part of mixed farming system that is typical of small holder agricultural systems. Dairy animals directly provide valuable inputs into our farm production activities and they function as important converters of crop residues into essential nutrition for whole community, especially for growing children with immediate beneficial impact on quality of life. India has emerged as the number one milk producers in the world and there are further great future scopes of progress in dairy sector in the rural development in the country. In the last 25 years, country has moved from a situation of scarcity and rationing of milk to a scenario which infuses considerable confidence in pushing ahead. Dairying is a sure path and future of our rural development.

Milk production in the country over the last two decade has increased by about 5 per cent per annum and the per capita availability of milk improved from 107 grams in 1970 to 214 grams at present. The per capita availability of milk has been increasing at a rate faster than the rate of increase of our population. Added to this growth rate in the fact that the cost of milk production in India is low as inputs are largely the by-products of the main agricultural activity. Not only is that most of the milk produced by labour houses in the margin. There is also a larger and rising domestic demand both for milk as well as for milk products. With an expanding middle income group, there even exist a market for value added products especially among quite a large segment of our population living in our cities and towns. There is also a possibility of converting surplus milk into unconventional milk products like casein and lactose which are not only import substitutes but also have a larger potential for export to developed countries. Modern scientific processing technology and managerial innovations being through about by our dairy managers are opening up before us opportunities of processing and marketing various types of value added milk and milk products.

Milk is consumed not only as a source of nutrition for our health but also, as products like, flavoured milk, various kinds of yoghurts, ice-cream, sweets and such other delicacies which would give larger price return to the industry and the milk producers as well. The emerging consumption pattern offered new challenges for the dairy industry, whereas various milk scheme has so far used to rationing the distribution of milk and milk products. Thus, the outlook of strategies is shifting from distribution to sales and marketing, opening up vast opportunities for the dairy industry.

The elasticity of demand for food grains being less than one, the impact of higher growth rate of income would result in shifts is household consumption. Milk and milk products would be one of the leading beneficiaries of increased consumption. In countries like China, where per capita income growth has been 9 per cent since last a decade, the demand for milk and meat recorded phenomenal increases. If income distribution does not become adverse to the poor, it is likely that the possibility of export of dairy products might hold higher potential than export of food grains. Since the milk production is different from many other agricultural per suits for reasons of extreme perishable nature of raw products, bulkiness and the biological process of continuous production, the gestation period which is a biological lag in output creates unique conditions and problems in handling milk production. Extreme perishability of milk and possibilities of contamination necessitates rapid movement of the product. The highly decentralized nature of the production asset, namely the milch cow and buffalo and the less than adequate development of physical infrastructure makes the desired speed of movement of milk difficult. The specially scattered nature of production brings up face to face with the problems of collection and transportation especially in linking remote villages with the market centres. Improvements in the organization to collect, process and distribute milk will reduce spoilage and dampen the seasonal variation in prices and there by expand the effective market for milk. Effort in breed improvement and genetic improvement through use of bio-technology and embryo transfer technology have kindled hopes of substantial increase in milk yields. Longer lactation periods and shorter calving intervals, organized milk collection, establishment of chilling centres, availability of laboratory equipments and others measures have great potential to reduce our milk procurement time and enhance the life of raw milk thereby minimizing the wastage.

The dairy processing industry itself undergoing in process of diversification and expansion, the structure of dairy industry is under a change consequent to the de-licensing of the industry and the economic liberalization process. The role of the private sector is expanding especially in those areas where there is a marketable surplus beyond the carrying capacity of the existing cooperative infrastructure. De-licensing has brought in many new players in the field and is leading to expansion of capacities licensed in the past. There are proposals from the private industry to enter into manufacture of industrial products like lactose and casein. Some of them have already entered into collaboration with foreign companies. The base of the industry is thus expanding.

The above process of expansion of the industry and the diversification of the structural industry needs to he matched with the corresponding increase in the production base, namely the production of raw milk. The dairy farming will remain substantially as an adjunct to agriculture sustained largely on crop-by-products and residues. The lack of empirical data makes it difficult to correctly assess the size class of producers. However, the size distribution of dairy farming may also need to undergo some change towards larger herd sizes. Dairy farming has to emerge as an economic enterprise in itself. Even while the preponderance of the total milch population may continue to be an integral adjunct of the agricultural production process, the growth in herd size almost is natural out growth of advancement in dairying as has been experience of the developed countries. Such a growth will promote generation of employment for educated youth and can attract them into dairy farming as a profession.

Anand pattern and Operation Flood ventured into hither to un-chartered quarters. It is not that the cooperative form of association was not unknown, but it remained an ideology and some one had to come forward with the tools for translating it into action. Dr. Kurien performed the role of the messiah and demonstrated how a good ideology holding in it the basic tenets of democracy would also be an equally good and powerful method of action. Thus, an ideological package was transformed into a meaningful package for running an economic enterprise, namely a cooperative society. The commitment to sustain the cooperative as a business enterprise, while at the same time ensuring its democratic functioning and member participation gave us a shining example of how small individual efforts could through a medium of cooperative can become the ideal form of economic enterprise for rural development. The fact that the Anand model can be replicated has also been fairly well established. The model has given us the strength to induct high yielding milch animals among our small, marginal farmers and to organize processing and marketing of milk output so created, thereby building a vibrant and vertically integrated cooperative structure that can withstand the fluctuations of the market.

There are vast areas still uncovered by the meaningful program of increase in productivity and organized efforts for opening up market channels. India has extensive geographical terrains coming under wide range of agro-climatic conditions with rainfall ranging from say 100 mm to as much as 4000 mm. A large part of our agriculture development is confined to the irrigated areas which constitute some 30 per cent of our total cultivated land. Innovative methods of rain harvesting and moisture retention in the soil are slowly giving rise to a sustainable scenario of rain-fed agriculture.

Fodder cultivation and grassland development will be absorbed as important components of development of a water-shed. A lot of animal resource development has taken place in irrigated agriculture and, therefore, potential areas for future perhaps may be the lower rainfall areas where dairying will also serve drought proofing. Thus economic development comes through being together. Can there be a better togetherness than being together with the economic prosperity of small farmers, marginal farmers and landless labourers. If they can produce 60 per cent of the milk output, by 75 per cent of membership of milk producers' cooperatives, this togetherness deserves to be strengthened.

International Scenario

Till now the increase in milk production has been attained not because of the enhanced productivity of the animals, but due to the increase in number of animals. Also, the increase in the proportion of buffalo's milk as compared to the cow's milk has contributed to the overall enhanced milk production. The dominance of buffalo milk in our total milk production has put India on altogether different platform in the international context, because the international areas are characterized by the dominance of cow's milk. There is greater flow of milk from the rural area to the urban areas, because of the urban-rural income disparities. The law of gravitation would show that since the urban income is growing as a faster pace than the rural income, therefore, the greater flow of milk is taking place from the rural areas to the urban areas. As a result, there is aggregate demand which is kept unfulfilled in the rural areas primarily because of the price incentives offered for supplying the milk to urban areas.

To explain this fact an example can be taken into account, that in Haryana which is a smaller state and where the cooperative dairy federation operates at much more shoe-string procurement and capacity utilization, 45 dairy plants have come up in the past decade. And all them have the urban markets of Delhi and Haryana. Haryana is one of the fastest growing urban conglomerates. Now this is the scenario number one.

Scenario number two which can again be painted clue from the rural urban dichotomy is that now the processing and distribution dimensions are going to play dominant roles, the dairy plants are divided into three sectors–First is a rural dairy which is predominately engaged in chilling operations, collection and them passing on to the feeder balancing plant or to a feeder plant. Second is the Union Dairy, and the third is the one which would predominantly be a processing dairy plant which is involved in manufacturing certain milk products. Now that is where the challenges lie before the Indian dairy sectors to compete the international arena. Here the distribution is one dimension which would have a predominant weight-age in our decision. Now the only way that can manage is by increasing the price to the consumer and reducing the share of the producer. That would be a challenge which is going to face and we have seen the consequences as problems in terms of capacity utilization and the whole chain will be disturbed.

Financing and Viability of Dairy Industry

There is need for the dairy industry to come out of legacy of the past and carve out its role in an environment of economic liberalization, policy deregulation and do away with the superfluous taxation, and legislation that were restricting the growth and development of the private and the co-operative dairy sectors particularly cooperative act. The industrial deregulation and opening up of the economic the procedure for appraisal monitoring and financing of all projects, including the dairy, are gradually becoming together and the customers who would tend to present proposals that are not viable would not find favour with the financers as the cost of capital is becoming expensive and therefore, it is necessary for the dairy industry to improve efficiency of utilization of all resources including money, machine and manpower. Further, it is urged for adoption of a new National Dairy Policy in which the private sector should be given adequate representation in participating and formulating such policies. The dairy industry to become financially viable it was necessary to improve the efficiency of all systems *viz.* milk production and collection, processing and product manufacture, marketing and distribution, and above all the manpower training institutions so as to produce skilled manpower and specialist in dairy industry.

Conclusion

Indian Dairy Industry is matchless in the world. It is emerging as a sunrise industry growing at a snail's pace unlike poultry industry and India is the world's largest milk producers with its lower per animal productivity till date. India produces 16 per cent the world milk production, under 20 per cent of which is processed. A unique feature of this industry is that over 50 per cent of the total milk production comes from the buffalo. It is unique in the sense that no where outside the South Asian region buffaloes are used as milch animals.

Buffaloes are India's milking machine that constitute less than 40 per cent of the bovine population, but contribute more than half of the total milk production. Albeit India has the largest cattle population most of them are non-descript animals, very poor in milk production. For a country like India having a population of 1.2 billion and availability per capita availability of milk would be woefully low-at 220 g. per day when it fulfills, the projected output of 9.0 million tones. Available data indicate that the average milk intake is as paltry as 20 g/day in eastern region as against a high of 400 g/day in the north India. In northern belt, majority of the people are lacto vegetarians, they compensate the non consumption of chicken, eggs, meat and even fish through milk consumption. Milk production of northern belt is much higher than eastern and south eastern region.

The dairy industry is much older than poultry sector and it is a rural based activity spread widely across the country in some 70 million households rearing one or two cows/buffaloes. Not with standing the poor productivity of native cattle, India has emerged as the world's number one milk producing country. The average lactational milk yield of cows and buffaloes is 445 kg and 911 kg, respectively. So, Dairying has now emerged as an important income generating and a source of employment opportunities in rural areas.

Thus the dairy industry has developed enormously with the milk production more than trebling in the last few decades. Nevertheless stile there need a long way to go to achieve the true potential that the country has. Generation of new technologies for landless group of livestock owners, an efficient inputs supply and services and linkage between different channels of production and marketing management systems are however urgently required in the years to come.

India has a great potential to dairy industry with its products exports. India's low cost milk will be price competitive in milk deficit countries *viz.* surrounding India. Asia is the world's fastest growing market for dairy products. According to an F.A.O. expert, "India is the slumbering giant of the international dairy trade. India has the largest bovine population 206 million cattle and 814 million buffaloes which indicate the bright future of dairy industry".

Chapter 4

Traditional Dairy Products: Some Aspects

G.S. Rajorhia
(Ex-Principal Scientist NDRI)
154 Sector-7, Urban State Karnal – 132 001

Introduction

Nearly half of the 90 million tones of milk produced in India is utilized for the manufacture of a variety of traditional milk products. Most of these products are produced on a small scale either in villages or by *halwais* in the cities and towns. Production centers are located in various pockets near big cities. The studies on the cost of production of some selected traditional products at these centers have shown good profitability. Besides being lucrative, the technologies used are simple with low operating and over head costs as compared with Western or European style products. A number of technological and social benefits can be attributed to indigenous milk products.

The art of preparing sweets from surplus milk was developed centuries ago. Lack of cooling facilities to keep liquid milk fresh in warm climate resulted in the diversion of milk for the preparation of indigenous milk products with comparatively longer shelf life. Milk food delicacies are value added products generating high profit. The demand for milk sweets is influenced by the nutritional and social value attached to each of them. In view of the growing awareness towards the safety aspects of milk based sweets in Indian cities and towns, the consumer shall prefer to buy these products from the organized establishments with a given brand. For this reason, some dairy plants are seeking information on technology of indigenous sweets.

The organized dairies at present handle and process about 12-15 per cent of the total milk produced in India. Collection of the entire surplus milk from about 5.8 lakh villages and its processing would require huge capital investment on equipment, building and other infra structure. The milk supplies in the cities and towns become very expensive as a result of high cost of procurement, processing and packaging and accompanying sourage losses. The large size dairy plants usually work on low throughput. Conversion of milk into indigenous milk products in and around milk production areas would be least expensive and more remunerative because of the simple technologies, use of low grade energy sources, inexpensive infra structure, local availability of equipment and low production costs as compared with other system of dairy processing. Indigenous milk products offer several technological and social benefits, generate employment, increase income and improve the economic status of the masses. Although authentic statistical data are not available, it may be estimated that about 45 million tons of milk is converted into traditional products annually in India.

Traditional Milk Products in Ancient India

Milk and milk products formed the principal ingredient of food of Vedic Indians. Khoa sweets have been the item of choice for centuries in India. In early Buddhist and Jain works, there is a mention that the sweets were prepared from inspissated milk named as *Sihakesara* and *Morandeka*. It appears that the rich used to enjoy such delicacies at the end of their meals. Buddha allowed his followers to take some sweets for journeys on routes where it was difficult to get food stuffs. In the Maurya period the sweets were prepared from concentrated milk with the addition of honey, jaggery or sugar. In the post–Gupta period (AD 750 to 1200), milk was used in various forms . When half of the quantity was evaporated, it was drunk. In case it was reduced to one third of the original quantity, it became a dish that was appreciated. When one-sixth of the original quantity remained, it was used for preparing sweets and when only one-eighth remained, it was called 'Sarkara' (powder). The ancient medical literature clearly states that the physical and mental happiness of individual depended on the food they took.

The art of preparing sweets from surplus milk was developed centuries ago. Lack of cooling facilities for liquid milk especially in warm climate resulted in the diversion of milk for the preparation of indigenous milk products with a view to achieve longer shelf-life.

Economic Importance of Indigenous Products

Milk food delicacies are value added products generating high profits. The demand for milk sweets is influenced by the nutritional and social value attached to each of them. In view of the growing awareness towards the safety aspects of milk based sweets in Indian cities and towns, the consumers would prefer to buy these products from the organized establishment. For this reason, many dairy plants are demanding information on technology of indigenous milk sweets. A critical review on the subject highlighting the weaknesses of the available technologies and the need for developing commercial production systems for each of the products will be presented.

Range of Indigenous Milk Products

Tables 4.1 and 4.2 represent the kind of traditional milk products which are prepared in India. This list is not exhaustive as many region specific products are made at different place and they do not find place here.

Table 4.1: Common Traditional Indian Milk Products

Process	*Product*	*Western Counterpart*
Concentration	Khoa	Evaporated milk
By evaporation	Basundi	Sweetened condensed milk
	Rabri	Sweetened clotted cream
	Khurchan	
	Malai	
Coagulation by heat and acid	Chhana	Direct acidified cheese
	Paneer	Latin American white cheese
Fermentation with lactic acid bacteria	Dahi	Plain yoghurt
	Payodhi	Sweetened-Yoghurt

Contd...

Table 4.1–Contd...

Process	*Product*	*Western Counterpart*
	Misti beverage	Concentrated and caramelized sweet yoghurt
	Chakka	Drained yoghurt
	Srikhand	–
	Srikhand wadi	–
	Lassi, Chhach	Stirred yoghurt
Phase inversion of fat	Makkhan	Butter
	Malai/loni	
Butter classification	Ghee	Butter oil
Freezing	Kulfi	Ice-cream and other frozen
	Kulfa	deserts
	Malai-ka-baraf	

The list of most common milk sweets and novelties prepared by combining one or more of the above indigenous milk products with non-dairy ingredients is given in Table 4.2.

Table 4.2: Some Milk Based Delicacies

Khoa Based	*Chhana Based*	*Cereal Based*
Variety of Burfi	Rasogolla and varieties	Kheer
Peda	Sandesh and varieties	Paysam
Kalakand	Pantoa, Lady Caning	Phirni
Gujhia	Raso-Malai	Mihidana
Palangtod	Cham-Cham	Kadhi
Malai-Laddoo	Ksheer-Sagar	Ras-kheer
Kunda/Dudh pak	Rajbhog	
Milk-Cake	Sita-bhog	
Gulabjamun	Chhana-murki	
Kalajam		
Carrrot-Halwa	Chhana-kheer (porridge)	
Gourd-Halwa	Lalmohan	
Malpooa	Totapuri	
	Chitrakut	

Present Status of Indigenous Products

Variations in Composition

The small traders and Halwais, except for about 20 per cent of total ghee contributed by the organized dairy plants, produce the indigenous milk products. The traditional methods are good for cottage scale, but do not permit efficient utilization of fuel. These processes are labour intensive. The quality of products varies in respect of moisture, texture and flavour. The quality of milk used, degree of heat treatment, manner of handling of products and duration and condition of storage lead to variations in the chemical composition, beyond normal statistical limits near two batches of same products prepared at the shop would match in proximate principle, not to speak about the large variation between the shops, standardization of techniques for uniform quality is necessary.

Unhygienic Conditions

The unhygienic conditions at the production unit lead to contamination of products with different types of microorganisms leading to a low shelf life of the base products. This problem gets aggravated during summer in respect of composition, microbiological quality, sensory properties and shelf life. A variety of micro-organisms comprising of millions of *mesophiles*, acid producers, spore formers, *coliforms*, *staphylococci*, yeast and moulds per gm can be found in the products.

Inadequate Packaging

Most of the products are sold in the market without proper packaging and unduly exposing them to atmospheric contamination. It is, therefore, not feasible to lay down any microbiological standards for these products.

Sensory Quality Differences

Ghee is the most important indigenous milk product utilizing nearly 29 per cent of total milk. Although it is a long life product, it has typical sensory problems. There are examples of market ghee being adulterated with vegetable fats, lard, tallow, etc. According to a survey of the leading ghee trading centers in the Northern India, it was observed that even the most selling branded ghee were deficient in one way or the other. The flavour defects reported by the trended panel of judges in the ghee samples collected from the market are given in Table 4.3.

Table 4.3: Sensory Defects in Market Samples of Ghee

Total Samples	96
Oxidized flavour	29
Rancid flavour	10
Smoky flavour	12
Fruity flavour	02
Burn/Cooked flavour	18
Bland or lacking flavour	09
Curdy/acidic flavour	06
Good normal flavour	10

Texture defects were also observed in the market samples of ghee the peroxide value in the above samples ranged between 0.0 and 12.72 as meq. of oxygen per kg of fat and 0.04 to 2.0 per cent moisture. Other constants were within the prescribed limits.

Modernization and Upgradation of Technologies

Industrial methods have now been developed for mechanized production, packaging and storage of khoa, chhana, ghee, dahi, gulabjamun, burfi, peda, rasogolla, kalakand and shrikhand. These products are being produced regularly on a factory scale. Many of these products are stored and transported in refrigerated cabinet and packaged in food grade plastic containers.

The production of traditional milk products by using modern methods would promote the utilization of larger quantities of milk during the flush season and help in stabilization of milk prices. With the help of recombined milk technology, traditional milk products can also be produced in lean months and in the milk deficient zones. Technology for recombined milk products is available for industrial application.

The recent developments in convenience traditional milk products category like khoa powder, rasogolla, mix powder, gulabjamun mix powder, kulfi mix powder, chhana powder, shrikhand powder, etc., with prolonged shelf-life and high consumer acceptability have supported the modernization efforts of this sector to a great extent.

Marketing of Indigenous Products

The demand for traditional products far exceeds that of western type dairy products. The increasing requirements for these products present a great opportunity for organized dairies to modernize production. However, most of these products because of high water activity, are perishable at room temperature. A provision of cold chain will, therefore, be necessary to preserve the traditional products for long periods. Maintenance of uniform quality throughout will promote marketing. Milk sweets sold by *halwais* at exorbitant rates are reported to carry abnormally high microbial load besides being substandard in chemical and textural properties. Since the children, are more prone to diseases and they are biggest consumer of milk based sweets, this reason alone should greatly favour the manufacture and marketing of indigenous milk products through organized channels. Effective packaging is yet another consideration for good marketing strategies. The products manufactured in the area of plentiful milk availability, especially during flush season, can be marketed in the places of milk scarcity and in urban areas, where milk is a scarce product and expensive because of high transportation cost. Milk sweets with well defined quality characteristics and packaged in attractive containers can be marketed at parlours, milk booths, hostels and restaurants, departmental canteens, supermarkets, airlines and railways catering, hostels and other points of strategic nature.

Future Directions

The need for up grading the technology for production, packaging, preservation and distribution of indigenous milk products is greatly realized now. Attempts to develop any large scale processes for milk based sweets must aim at using the already existing process in food plant rather than relying on a new plant design and fabrication which carry limitation of available engineering skill, high cost and time loss. In view of the growing market demands for indigenous milk products and their derivatives, there is a great scope for adopting the existing unit process technologies for industrial production. Many new processes such as membrane technology for concentration of milk have been recommended in the recent past. There is a considerable scope for using process for conserving regional and seasonal milk surpluses. The chemical and microbiological changes during storage and transportation diminish both the nutritional quality and consumer's acceptance of milk sweets. High ambient temperatures and relative humidity greatly contribute to the spoilage problem. A good possibility exists for improving the shelf-life of milk sweets by employing few of the selected approaches such as modified atmospheric or vacuum packaging, use of humectants and natural preservatives and adherence to good manufacturing practices. The quality of raw milk, of course, is of paramount importance for producing high-grade milk products. Enforcement of standard sanitary practices at all stages of products handling, packaging and distribution will boost marketing and lead to heavy dividends. Quality assurance in terms of food safety and freedom from contamination is desired to be achieved. Risk assessment during milk production and processing must receive priority is dairy research and education in India to become globally competitive.

Chapter 5

Buffalo Production in India

B.L. Saraswat

Department of Animal Husbandry and Dairying
Udai Pratap Autonomous College, Varanasi – 221 002, U.P.

The domestic or water buffalo (*Bubalus bubalis*) commonly known as an 'Asian Animal' is the dairy, draught and meat animal of Asia. The Indian subcontinent is the home tract of the world's dairy buffaloes. Buffalo is destined to play an important role in National and International economy for a long time.

Why Buffalo Production

- ☆ In India about 55 per cent milk is shared by buffaloes.
- ☆ Buffalo milk contains 30 per cent more total solids and bigger size of fat globules as compared to Indian cow milk.
- ☆ Buffalo milk contains about 15 per cent, 55 per cent, 7 per cent and 30 per cent more amount of protein, fat, lactose and energy respectively than the Indian cow milk.
- ☆ The digestibility of buffalo milk has been estimated to be 88.86 per cent and biological value of its protein is 84.04 per cent.
- ☆ Buffalo milk is a better source of Ca, P and Vitamins as compared to Indian cow milk.
- ☆ Buffalo milk contains natural antioxidant 'Tocopherol' which helps in the preservation process.

- ☆ Buffalo milk has high concentration of fat but low cholesterol content.
- ☆ Buffalo milk protein contains sufficient amount of essential amino acids, which are required for body growth as well for proper growth of the brain.
- ☆ Buffalo meat has higher palatability, less cholesterol and higher protein content than cattle meat.
- ☆ The dressing percentage of male buffalo calves has been reported to be in the range of 55.4 to 59 per cent on moderate diet and as high as 64 per cent on low protein and optimum energy diet.
- ☆ Male buffaloes are considered better for draught purpose.
- ☆ Buffaloes utilize low quality roughages and convert them in high quality and high energy milk.
- ☆ Buffalo is slow, ponderous and docile animal.
- ☆ It has remarkable capacity to adopt itself in the most adverse climatic conditions. The best milch buffaloes are found in areas where temperature goes as high as 40°C to as low as 4°C.
- ☆ Buffaloes are comparatively less prone to mastitis as they have tighter teat sphincters as compared to cows.
- ☆ Buffaloes have longer teats, which facilitate the milking process.

Contribution to National Economy

In India buffaloes contribute about 55 per cent of the total milk produced with the increment rate of about 5 per cent annually. Since buffalo milk has high fat content it is sold on high rate and always 2-4 rupees per litre higher as compared to cow milk. Buffaloes with an increment rate of 3.5 per cent contribute more than 20 per cent of total meat produced annually. Apart from the milk and meat, eight million buffalo bullocks are indispensable to agricultural operations. Leather obtained from buffaloes is used in the preparation of important goods and provides export opportunity also.

Buffalo Population and Distribution

It has been estimated that there are about 163 million buffaloes in the world, which is about 10 per cent of the world's bovine population. Of these 97 per cent are found in Asian Countries.

As such largest population of buffaloes is in Asia and Pacific countries with an annual rate of increase by 1.7 per cent as compared to 0.7 per cent in cattle. At present more than 92 million buffaloes are found in India of which about 50 millions are milch buffaloes, 18 millions are used in farm activities, young stock and breeding bulls and about 24 millions females are below three years. China has the world's second largest buffalo population.

The domesticated buffaloes may be classified into two main categories:

1. *Reverine buffalo*: Having 25 pairs of chromosomes and is mainly milk producing buffalo. Reverine buffaloes are found in West Asia mainly in India, Pakistan, Bangladesh, Nepal and Sri Lanka.
2. *Swamp buffalo*: Having 24 pairs of chromosomes and is found in East Asia mainly in Philippines, Thailand, Malaysia, Indonesia, Taiwan, Singapore and China.

Reverine buffaloes of Indian subcontinent prefer the clear water of streams and pools bunching close together but the swamp buffalo of East Asia likes to wallow in slime, in a mud hole accommodating one animal or very few animals only.

Buffalo Breeds in India

Only 15-20 per cent of total buffaloes population are of well defined breeds and rest 80-85 per cent are classified as non-descript. There are eight main breeds in India *viz.* Murrah, Bhadawari, Mehsana, Surti, Nilli Ravi, Jaffarabadi, Nagpuri and Tarai.

Economic Traits of Buffalo

Reproductive traits

Age of maturity	: 30 months
Oestrus cycle	: Av. 21 days
Oestrus period	: 18-24 hours
Ovulation time	: Av. 11 hours after cessation of oestrus
Optimum time for insemination	: 12 hours after the outset of oestrus.

Age of first calving	: 39-56 months
Gestation period	: 310 days
Intercalving period	: 390-420 days
Time for expulsion of placenta	: 6 hours
Body weight of calf at birth	: 30-40 Kg

Productive Traits

Lactation length	: 300-305 days
Lactation yield	: 1800-2200 lit.
Av. daily milk yield	: 7.0 lit.
Dry period	: 65-90 days

Physiological Parameters

Body temperature	: 99-102°F
Pulse rate	: 40-45 per min.
Respiration rate	: 16-18 per min.

Programmes for Rapid Genetic Improvement

As previously stated 80 to 85 per cent buffaloes are non-descript in India having low genetic potential. Therefore, scientists have diverted their attention towards the up gradation of low productive herd of buffalo. In this direction certain biotechnological practices are on progress:

1. Synchronization of oestrous.
2. Super ovulation or multiple ovulations.
3. Embryo Transfer Technology.
4. Genetic cloning.

Synchronization of Oestrous

- ☆ High yielding buffaloes do not return to oestrus after calving for some times they take 90 days or even more.
- ☆ High yielding buffaloes generally have silent heat.

To overcome all these problems it is better to get buffaloes synchronized for oestrus and inseminate them at an early date. For this purpose, artificially prepared Progesterone hormone

(Progestogens) is used. Commonly available Progesterone preparations and their doses are as below:

1. MAP (Methyl acetoxy progesterone) – 10 mg/day (less potent)
2. CAP (6 chlor-6 dehydro 17-acetoxy progesterone) or Chlormedinone- 10 mg/day
3. MGA (Melengesterol acetate) – Only 1mg/day (highly potent).

All these preparations are given orally for 14 to 18 days. Progesterone can be administered daily intramuscular injection of 25 mg for 7 days for desired effect.

Super Ovulation or Multiple Ovulations

To obtain more than one ovum at a time from high yielding buffaloes this practice is followed. A dose of 4 – 5 mg FSH is administered two times a day upto 4-5 days after that an injection of prostenglandin F_2 is administered between the oestrus cycle or after 9-10 days of heat period. Now the ovary will produce more than one ovum, which can be transplanted in the uterus of other animal.

Embryo Transfer Technology

During this process, the embryo of a good, healthy and high yielding buffalo (Donor) is transferred to the other buffalo (Recipient). This is a process by which more number of best calves may be obtained from a female within short period.

Genetic Cloning

To change in genetic structure for producing desirable characteristics with the help Molecular Biology is called 'Genetic Cloning'. In this process gene of an animal is conjugated with gene structure of other animal and has 3 steps:

- ☆ To separate the gene from cell
- ☆ To enter the gene in another life
- ☆ To ascertain that recipient has accepted this gene and its characteristics have been adopted.

Chapter 6

Quality Control in Milk Products

Rajendra K. Pandey
Department of A.H. and Dairying,
B.H.U., Varanasi

Indian has made great stride in milk production and presently ranks first in the world. Quality milk production will be a major challenge for the 21st century. It is to be viewed in the back drop of deteriorating milk quality standards in the country. We have achieved tremendous improvement in milk production but the quality aspect have been most neglected, leading to poor quality milk and milk products. Moreover, we are to face a more open and liberal market both domestic and abroad. With this changing scenario, we have to gear ourselves towards better quality of milk to meet the demands of the market.

Cross breeding programme and changes in feeding pattern of milch animals, the milk of some animals does not meet the legal PFA standards. This has resulted in the adulteration of milk even with harmful substances to raise requirements. In the recent past, due to liberalization policy of Government of India, there is tremendous mushrooming of dairy plants which has resulted in scarcity of natural milks especially during summer season. The milk procurement also starts dealing during the lean season. This is the period when unscrupulous lements try to harvest the gain by adulterating the natural milk using all type of pond water, salt sugar, wheat flour, baking soda, washing soda, caustic soda, urea, hydrogen peroxide, formaline and different vegetable oil and animal body fat have been known to be in use depending on need and situation. Some of the adulterants are used for increasing the volume of milk, while others are used for preserving and extending the shelf life and to meet the legal standards of milk. The phenomenon of adulteration goes on changing in terms of ingredients to make maximum profit. In November 1994 reports appeared in different news papers and magazines indicating that in several parts of the country, the milk vendors are selling synthetic milk which is prepared using harmful chemicals. The term synthetic milk is misnomer and is probably coined by the media person for adulterated milk. From synthetic milk one would normally understand a product analogous to natural milk in its physical, chemical and nutritional properties. But there is no similarity between the two. Who can synthesize milk other than milch animals? Synthetic milk is gross misuse of knowledge of chemistry *i.e.* process of saponification. The basic ingredients used in the manufacture of synthetic milk are caustic soda, water, refined vegetable oils, animal body fats, urea, detergents, sugar, salt and skim milk powder. Detergents increase viscosity of the solution and vegetable oil prevent frothing in the solution. Soap sopanifies the oil into an emulsion which liberates free fatty acids into the milk.

Adjustment of non fat solids in milk is done and acidity is neutralized by using caustic soda. Lactometer is used to adjust specific gravity of synthetic milk equal to that of natural milk.

In the past WTO regime of liberalized trade, regulatory aspects of food standards are crucial to words consolidating the benefits of new free trade agreement. Dairy industry will be able to amass possible benefit by being a global contender in trade only after its products are at par with those of the rest of world. The Indian food industry will sustain only if it is able to provide a competitive edge vis-a-vis price and quality.

Quality

Quality has been defined as "The totality of characteristics of an entity that bears on its ability to satisfy stated and implied needs". The importance of quality can best be explained in the words of sueo yamaguchi, consultant. J I P M "Quality is not some thing to be discussed in Japan; it is not something you talk about. It should be there, that's all that is how it should be in India too".

There was a time when people used to stand in long queues to get their daily quota of milk. Quality was not critical. Today the scene is different; no one sees long queues at milk booths. If customers are not happy with supplier, they go to another. Dairies have understood that quality products and services are the key factors for winning over customer.

Emphasis on Quality Control

Strict system of quality control is needed in order to provide consumers milk and milk products with high quality. The system begins with production and procurement of milk. Milk should be tested at the processing centre for fat, S. N. F., acidity, bacteriological quality and presence of preservatives or neutralizers which are not permitted by law. After processing the milk should again be checked for bacteriological staying qualities, as well as fat and S. N. F. content in order to ensure that it meets the rigid standards. Spot checks should also be made at retail outlet's to ensure that the required quality standards are met. Besides product control strict control should be kept on sanitation, both at the plant and retail outlets. Besides chemical and bacteriological tests for products, processing equipment should also be subjected to regular bacteriological tests. Water in dairy should be softened, chlorinated and regularly tested to maintain highest standards. Quality of water used at booths and insulated containers should also be checked.

International and Indian Standards for Dairy Products

To keep pace with newer scientific and technological developments, the revision of standards has been entrusted to the Codex Committee on Milk and Milk Products (CCMMP), established in 1993. During its several deliberations, 11 standards have been revised and adopted by the CAC. Owing to interventions by the Indian delegation, certain alterations have been made in some standards. A few pertain to the definition of milk, inclusion of ghee in the standard Milk Fat Product, incorporation of BHA as an antioxidant and removing the restriction of permitting only cow milk as a raw material, particularly for cheese. Due to continued efforts by India at the IDF as well as CCMMP, a scientific proposal illustrating the suitability of buffalo milk for cheese making was prepared, thus relaxing the earlier restriction.

There are certain areas where discrepancies still exist in Indian and international standards. In India, a level of 0.2 ppm of lead in milk is considered safe, whereas international requirements are fixed at 0.02 ppm. The level of lead in butter (maximum of 0.05 mg/kg) was not acceptable to the Indian delegation, but was adopted by the CAC as it had been already endorsed by the Codex Committee on Food Additives and Contaminants (CCFAC). Differences between existing Codex and PFA standards of pesticide residues in milk have been documented by Varshney (2000). The differences in some of the quality standards for ghee are also listed. There are CAC guidelines for the use of lactoperoxidase system for the preservation of milk, but the LP – system is not permitted in India. Similarly, using decolorized buffalo milk is permitted in several nations, but not India.

International standards on quality and Food Safety Management Systems

World is witnessing technological breakthroughs and emerging global markets. Global standardization and certification systems encapsulate these developments and provide tools to facilitate international transactions of goods and services. Thus standardization and quality systems have become sin – qua – non for development of the national economy all over the world. It is in this context that ISO 9000 quality management systems standards brought out by International Organization for Standardization (ISO) and Food Safety Management Systems (HACCP) standards brought out by the Codex Alimentarius Commission (CAC) have assumed greater importance for achieving the objective of facilitating global trade for safe and whole some foods.

HACCP for Quality Management in Dairy Trade

Hazard analysis and critical control point the international food handling safety standards ensure cleanliness, hygiene and sanitation in handling milk and milk product and give an assurance of safety to consumers. Indian is signatory to W. T. O. agreement and have understood its significance and started the process of educating

application of international standards, international certification systems and bringing transparency in the–promulganation of national regulatory requirements as means of facilitating international trade of food. SPS is particularly applicable to food industry and adoption of HACCP standard under the agreement makes it mandatory for the food industry to apply food safety management systems for facilitating international trade.

TBT

It covers all the types of standards including aspect of food standards (except requirements related to sanitary and phytosanitary measure) and include large number of measures designed to protect the consumer against deception and economic fraud. Food standards it covers relate to quality requirements, labeling and methods of analysis. The agreement provides all the technical standards and regulation must have a legitimate propose and that the impact or cost of implementing the standard must be proportional to the trade. It should also facilitate trade. Under this agreement, each country has to operate an enquiry point, which is able to answer all reasonable enquiries from member countries and interested parties on technical regulations, standards and conformity assessment systems.

Sanitary and Phyto Sanitary Agreement (SPS)

The SPS agreement requires that member countries base their national measure on international standard, guideline and other recommendation adopted by codex Alimertarius commission, a joint FAO and WHO commission, where they exist. As lodex standards have the full support of SPS agreement which advocates them as basis for all national standards, they will play a significant role in harmonization of national food safety standards with increasing international trade in food stuffs and specific requirement of SPS agreement there is a need for uniform guideline for inspection and certification systems in all countries. The SPS agreement requires that inspection and certification system should ensure transparency, equivalency and risk based systems.

Integration of ISO 9000 and HACCP

ISO 9000 encompasses all the activities of a dairy to ensure that it meets its quality objectives and HACCP system is directed towards ensuring safety of food. ISO 9000 and HACCP, one concerned with quality and another for food safety management systems respectively, have much in common. In fact quality in broader sense includes food safety.

Conclusion

Consumer trades are changing. Technological advances are taking place. Organic dairying is the emerging slogan. At this junction consumer satisfaction is the key force that will determine the future strategies with regard to quality and safety of food.

Chapter 7

Livestock Production Scenario of India

B.L. Saraswat

Department of Animal Husbandry and Dairying, Udai Pratap Autonomous College, Varanasi, U.P.

Livestock production in India is closely linked with agriculture and plays a prominent role in the rural economy in supplementing the income, particularly to the landless and small and marginal farmers. This sector also provides subsidiary occupation in semi urban areas and for the people living in hilly and tribal areas also. At present 70 per cent of Indians are dependent for their livelihood on agriculture. Thirty per cent of the total land holding are held of seventy per cent of small and marginal farmers who held 80 per cent of the total livestock in the country.

Contribution of National Economy

In India agriculture accounts for 27 per cent of GDP and livestock accounts for 26 per cent of agriculture GDP. The contribution of livestock to the National Economy is estimated to be about 70 per cent from milk and milk products, 11.2 per cent from meat and meat products, 8.8 per cent from poultry and poultry products, 7 pre cent dung for fuel and 3 per cent for wool, leather, bones etc.

According to provisional estimates of national accounts, the gross value of output for livestock sector during 1997-98 was about Rs. 111.40 billion as against the total value of output of Rs. 449.50 billion for the agriculture sector. This excludes the contribution of animal draught power. Export earning from the livestock sector and related product rose Rs. 1.925 billion in 1996-97 from Rs. 0.792 billion in 1988-89.

Employment Opportunities

The livestock sector also plays significant role in supplementing family income and generating employment besides providing cheap and nutritious food to millions of people. It has been estimated that livestock sector provides employment to 9.8 million on principle status and 8.6 million on subsidiary status.

Livestock Population

India ranks first among major livestock holding countries in the world and its livestock inventories about 500 million based on an annual growth rate of more than 1.5 per cent. India constitutes less than 3 per cent of the world total land area but it supports 56 per cent of the world's buffaloes, 16 per cent of the world's cattle, 18 per cent of world's goat, 5.4 per cent of the world's sheep and 1.7 per cent of the world's swine population (FAO 1998).

According to livestock census (1992) India has 210 million cattle, 92 million buffaloes, 120 million goats, 56 million sheep, 16 million pigs and 6 million camel, horses etc.

Indian Breeds of Livestock

At present India has 27 breeds of cattle, 8 breeds of buffaloes, 42 breeds of sheep, 20 breeds of goat, 7 breeds of camel, 8 breeds of horses, 17 breeds of poultry. Only 18 per cent of total stock of cattle and buffaloes are of well-defined breeds and rest 82 per cent are classified as non-descript animals.

Availability of Feed and Fodders

The total annual requirement for green fodder, dry fodder and concentrate is estimated as 761.53, 650.70 and 120.58 MT while their availability is 573.50, 387.86 and 42.98 MT respectively. Thus, there is a deficit of 25 per cent, 40 per cent and 65 per cent of green fodder, dry fodder and concentrate respectively. (*Ranjhan, 1997*).

The requirement of nutrients in terms of TDN and DCP have been estimated as 454.80 and 40.90 MT against the availability of only 253.90 and 22.0 MT respectively (*Jain et al., 1996*). Thus, there is a deficit of about 44 per cent and 46 per cent of TDN and DCP, respectively.

Milk Production and its Consumption

In 1998, India occupied first rank in the world regarding milk production with its output about 78 million tones which has been increased to the tune of 90 million tones in 2004. In which 55 per cent and 43 per cent is shared by buffaloes and cattle respectively. The per capita availability of milk in India is about 233 g/day or 85 kg/year but it varies widely in different regions of the country.

About 45.7 per cent of total milk produced is India is consumed as fluid annually and the rest 54.3 per cent is converted into the preparation of milk products like butter and ghee (34 per cent), dahi (6.9 per cent), khoa and condensed milk (6.5 per cent), milk powder (3.7 per cent), paneer, chhana and cheese 1.9 per cent and icecream, kulfi, cream and other 1.3 per cent.

Meat and Egg Production and their Consumption

In India, about 11.5 per cent of total income due the livestock production comes from meat and meat products. During 1996-97 total meat production was over 2.0 million tones possessing about 32 per cent meat of cow and buffalo, 40 per cent of goat and sheep and 11 per cent of pig, 15 per cent of poultry and 2 per cent meat of rabbit, deer, camel etc. The export of meat and meat products during 1996-97 was Rs. 0.693 million which accounted about 36 per cent of total exports from the livestock sector.

The egg production increased to the level of about 28.5 million during 1997-98 and currently India ranks 5th in egg production, 18th in broiler production and 22nd in poultry meat in the world. At present annual per capita availability of eggs and poultry meat in India is about 36 eggs and 850 g respectively. In India about 75 per cent of total consumption of meat and egg is performed by the people of urban area and 25 per cent of the rural area.

Wool and Pashmina Production

About 43 million wool is produced by sheep in India annually which has both coarse and fine fibres. In addition to this, Pashmina a very fine fibre is produced by a particular breed of goat in hilly tract of the country.

Inspite of milk, meat, egg and wool production, livestock also provides hides and skins, hooves, bone and hairs. These are used in the manufacture of important goods, which accounts a significant value and provides export opportunities.

Programs for the Development of Livestock Production

Various programs for the development of livestock production in the country have been launched time to time by the Govt. of India.

Programs	*Starting Year*
☆ Etawah Pilot Project (EPP)	1948
☆ Community Development Programs (CDP)	1952
☆ Key Village Scheme (KVS)	1955
☆ Intensive Cattle Development Projects (ICDP)	1965
☆ National Demonstration Programs (NDP)	1965
☆ Operational Research Projects (ORP)	1968
☆ Training and Visiting program (T&V)	1974
☆ Krishi Vigyan Kendra (KVK)	1975
☆ Lab to Land Program (LLP)	1979
☆ National Agricultural Research Project (NARP)	1979
☆ Krishi Gyan Kendra (KGK)	1980
☆ District Rural Development Agency (DRDA)	1980
☆ Livestock Insurance Scheme (LIS)	1993

Chapter 8

Use of Probiotics in the Feeding of Ruminants

B.L. Saraswat

Department of Animal Husbandry and Dairying
Udai Pratap Autonomous College, Varanasi – 221 002, U.P.

Nutritionists and microbiologists are facing a great challenge to optimize the productive efficiency of livestock by manipulating the microbial ecosystem present in the gastro-intestinal tract of the animals. A newly born calf of ruminant animal is completely sterile but when it comes in the contact of outer environment, its mother, other animals or feed, the calf is contaminated with the microbes, which colonize rapidly in the gastrointestinal tract. This microbial environment inside the gastrointestinal tract contributes to the overall performance of the animals by protecting them from various diseases.

A fully developed rumen has complex microbial eco-system, which enables the ruminants to utilize lignocellulosic feeds. Any disturbance in the microbial balance may affect the performance of the animals adversely and any improvement in the microbial environment may lead to the improvement in the productivity of the animal. Various attempts have been made in this direction by using chemicals like antibiotics, microbial growth factors, methane inhibitors, etc.

The usefulness of microbial feed additives came into light due to a belief that the longevity of Bulgarian peasants was related with large consumption of milk fermented with *Lactobacillus acidophilus.*

Probiotics may be defined as the organisms and substances that contribute to intestinal microbial balance. In Animal Nutrition, probiotics are viable microorganisms used as feed supplements, which lead to beneficial effects for the host animal.

Modification in population of rumen microbes by supplementing the feed with probiotics has been found to stimulate growth and production because of significant improvement in feed utilization. A wide range of microbial feed additives including bacterial culture; mixture of bacteria, yeast culture and fungi may be supplemented with the feed of ruminant animal. These probiotics affect directly microbial activities in the rumen which alter VFA production, stabilize rumen pH, increase propionate and reduce acetate, methane and ammonia production by increasing concentration of anaerobic and cellulolytic bacteria.

In the same way, 'Effective Microorganisms' (EM), consisting of an efficient culture of beneficial microorganisms, has also been tried on cattle, sheep and goats in abroad. The results of EM technology have been quite encouraging within a few weeks of its use. General health of animal was improved, bad odour and flies were controlled, severe wounds were healed in short period and the milk yield was increased by 13 per cent in cattle. Thus for increasing growth and production of ruminant animals dietary supplementation of probiotics has immense importance.

Types of Microbial Feed Additives

A. Fungal and Yeast Cultures

- ☆ *Aspergillus niger, A. orygae* (AO)
- ☆ *Saccharomyces cerevisiae* (YC)
- ☆ *Candida pintolopesii*

B. Bacterial Cultures

- ☆ *Lactobacillus acidophilus, L. brevis, L. delbreukii subsp. bulgaricus, L. casei, L. curvatus, L. cellobiosus, L. fermentum, L. lactis, L. plantarum, L. reuterii*

- ☆ *Bacteroides amylophilus, B. capillosus, B. bifidum, B. infantis, B. longum, B. thermophilum.*
- ☆ *Bacillus coagulans, B. lincheniformis, B. bifidum, B. pumilus, B. subtilis.*
- ☆ *Leuconostoc mementeroides.*
- ☆ *Pediococcus acidilacticii, P. cerevisiae, P. damnoses, P. pentocaceus.*
- ☆ *Propionibacterium freudenreichii, P. sharmani.*
- ☆ *Streptococcus cremoris* (renamed *Lactococcus lactis*), *S. diacetylactis, S. faecium, S. intermedius, S. lactis, S. thermophilus.*

Lactobacillus species are common probiotic products to human nutrition but in animal nutrition *Enterococcus* spp., *Saccharomyces* yeast and spore forming *Bacillus* spp. are most frequently used. Although 19 probiotics preparations with their mode of action have received temporary approval in the European union, which lead to beneficial effects.

Mode of Action

The beneficial effects of probiotics supplemented with ruminant feed have studied extensively with special reference to *lactic acid producing bacteria* and *yeast*. However, these two groups are quite different in their mode of action as the former group is primarily responsible to take care of health especially in young animals and the later one primarily responsible to take care of rumen function. In brief there are three main actions of probiotics in ruminant animals as below:

1. Decrease in viable count of undesirable microbes by
   (*a*) Production of antibacterial compound
   (*b*) Competition for adhesion site
   (*c*) Competition for nutrients.
2. Alteration of microbial metabolism by
   (*a*) Increased enzyme activity
   (*b*) Or decreased enzyme activity
3. Improved immune system by
   (*a*) Increased antibody activity
   (*b*) Increased macrophage activity

Action of Lactic Acid Producing Bacteria

- Lactic acid bacteria produce organic acids like acetic and lactic acids in large quantities, which result in lowering of pH. The organic acids are toxic for the undesirable micro flora in the intestine and this toxicity is increased many manifold at low pH.
- *Lactobacilli* colonize quickly in the gut to displace the coliforms and other pathogens from gastrointestinal tract, thus it reduces the incidence of diarrhea in claves and piglets.
- Various strains of *Lactobacilli* produce antibiotic like substances mainly *acidophilin, acidolin, lactobacillin, lactocidin* etc. which have been found effective against pathogens.
- Hydrogen peroxide produced by the *Lactobacilli* has strong germicidal activity at low pH.
- Most of the lactic acid producing bacteria have the ability to deconjugate bile acids to free from *i.e.* cholic acid, deoxycholic acid and taurocholic acid. These free acids are more inhibitory and are more effective against anaerobic and Gram positive aerobic bacteria.
- *Lactobacilli* have also been found to increase the activity of macrophages, lymphocytes and immunoglobulins in animal body.
- The microbial feed additives reduce the absorption of cholesterol in the intestine. Thus feeding of *Lactobacilli* decreases the level of cholesterol in blood.

Action of Fungal and Yeast Cultures

- Feeding of fungal cultures increases microbial protein synthesis and also changes the amino-acid profile stimulating the growth of specific bacteria.
- Esterase enzymes secreted by *Aspergillus oryzae* is considered as an important factor responsible for stimulation of fibre degradation.
- Feeding of yeast substantially increases the total viable count (TVC) in rumen (30 to 90 per cent) by creating more favourable environment for the growth of bacteria and altered rumen fermentation pattern.

- ✩ Increase in TVC results primarily by increase cellulolytic bacterial count, which enhances rate of cellulose fermentation.
- ✩ Yeast feeding in ruminants, stabilizes pH of the rumen thus favouring the growth of cellulolitic bacteria which are very sensitive towards pH.
- ✩ A large amount of oxygen enters in GI tract with intake of feed, which adversely affects the rumen microbes, which are strict anaerobes. By feeding yeast culture the 50 to 90 per cent oxygen is disappeared.
- ✩ Yeast cultures are rich in vitamins and minerals, which provide stimulatory effect to the cellulolytic bacteria.
- ✩ Yeast feeding altars fermentation pattern in rumen thus the digestibility of DM, ADF and NDF is increased.
- ✩ Yeast feeding increased intake and improved N, Zn and Fe metabolism.
- ✩ Glutamic acid produced by yeast is responsible for improvement in the taste of feedstuffs and a pleasant odour thus supplementation of yeast in the diet increases the palatability of feed.
- ✩ Since yeast has highly ionic cell wall, it is possible to remove a number of toxic factors entered in the gut through feed, which inhibit the growth of undesirable rumen microbes.

Selection of Probiotics

Microbial cultures should be selected having following characteristics:

1. A viable product can be formed at industrial scale.
2. It should be stable and viable during long storage and field conditions.
3. It should have capability to survive (if not grow) in the GI tract.
4. It must produce beneficial effect to host animal.

The efficiency of microbial additives can be tested using *in vitro* method. Viability is a major issue while selecting the probiotic preparations while preparing probiotic products, the microbes undergo various processing and may change its form. An efficient culture is, which can face these changes and can rejuvenate normally when suitable conditions are provided.

Chapter 9

Screening for Abiotic Stress Resistance

J.P. Srivastava

Professor and Head, Department of Plant Physiology
Institute of Agricultural Sciences, B.H.U., Varanasi – 221 002

The response of plants to the severities of their environment have occupied the attention of man long before the beginning of science of biology (Levitte, 1972). Plants that survive to these environments were termed as " hardy" and that do not were refered to as "tender". Biologist have coined the term "stress" for any environmental factor potentially unfavourable to living organisms. In mechanics term "stress" is well defined but such definition has not been developed in biology. Due to lack of an exact terminology in biological science, it leads to a lack of exact concept.

There are two main types of environmental stresses; biotic and physicochemical. The former belongs to field of pathology and ecology. The major physico-chemical stress, to which plants may be exposed are water deficit, flooding, temperature, radiation, chemical, wind, pressure, magnetic and sound stresses.

Broadly plants adopt avoidance or tolerance strategies to resist stress. Infact, each and every plant on this earth may experience stress of one type or another during its life time. Stresses are the major constraint limiting crop productivity (Buchanan *et al.*, 2000)

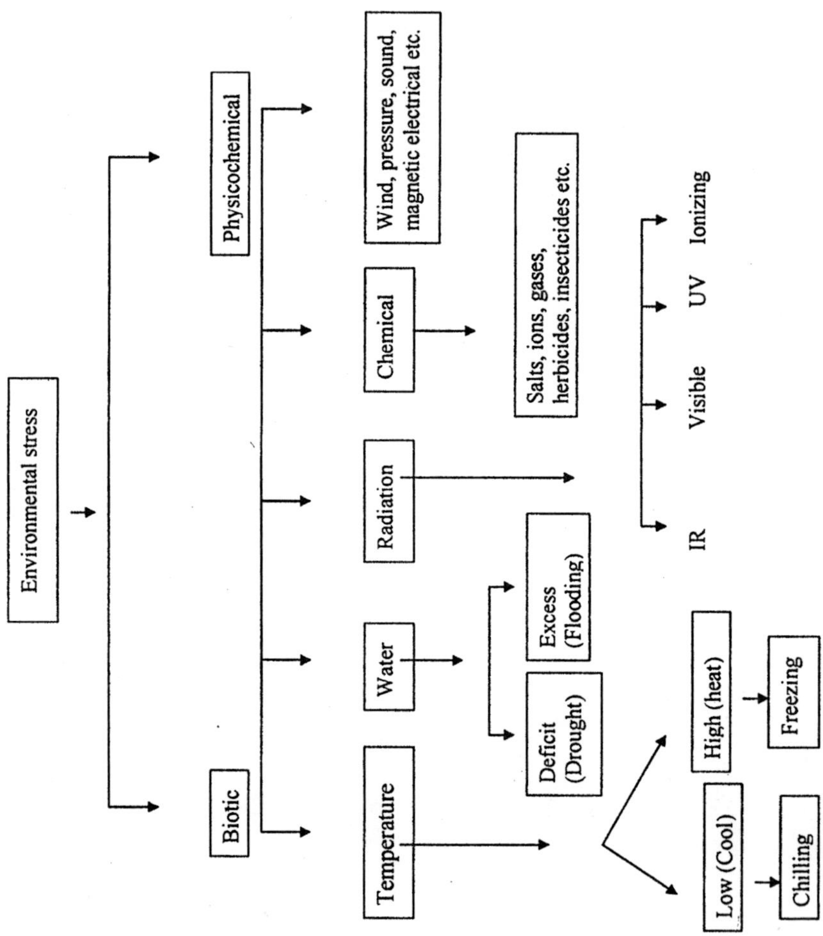

Figure 9.1: Kind of Environmental Stress which a Plant may Experience

and are often associated with food shortage and overall setback to developmental activities. Therefore, raising stress resistant crop may help to achieve stability in production and enhance the possibility of self reliance in food grains, vegetables, horticultural and forage crop production.

To achieve these objectives either of the following strategies are to be adopted for a given stress:

1. Identification of crops and their genotypes resistant to stress.
2. Identification of traits associated with stress resistance.
3. Development of plant type of desired characters by breeding, genetic engineering and/or biotechnological means.

In India major cropped area is under rainfed condition and crop productivity from such regions are poor. Therefore, further improvement in crop production is possible simply by selecting moisture stress resistant crop and their varieties and adopting suitable management practices. Here, I would like to explain the techniques to screen drought resistant crops and genotypes. Some of these techniques are:

Field Screening T

Crop is grown at different soil moisture regimes. Soil moisture is monitored at a definite depth. Crop performance in relation to soil moisture is evaluated.

Leaf Rolling

It is a common symptoms of deficit, with occurs due to loss of turgor. Leaf rolling decreases heat load on plants (Gates, 1968), decreases transpiration rate. Delay in leaf rolling may be an indication of turgor maintenance. In some of the plants this technique has been used for mass screening (O' Toole *et al.*, 1979).

Lime Source Gradient Technique

Genotypes are planted in rows or narrow long plots perpendicular to irrigation line source, and are exposed to the gradient of moisture stress and their performance is evaluated.

Leaf Firing

Extent of leaf firing as negatively correlated with stress yield in maize and sorghum. Leaf firing occurs due to cellular damage at elevated temperature on account of reduced transpiration rate.

Leaf Canopy Temperature Measurement

It is another parameter, which is dependent on transpiration. Leaf to air (Srivastava, 1989) and canopy to air (Idso *et al.*, 1981) temperature differences have been very effectively utilized to screen drought resistant genotypes.

Epicuticular Wax Content

In most of the crop plants nearly 20 per cent of total transpirational losses are through cuticular. Under moisture stress and xerophytes this value may exceed (Srivastava and Kumar, 1994). Thick deposition of epicuticular wax, a component of cuticle, reduces water loss and decreases leaf temperature by reflecting light radiation.

Root Studies

Root to shoot ratio has been considered as an screening criteria for a long time. Under field conditions it is difficult to study this parameter precisely due to interference with weed roots. It is advocated that instead of root to shoot ratio, root length density (cm root per cm^{-3} soil) should be taken the screening parameter.

Evaluation of Seedlings in Different Osmoticum

It has been a long practice to grow the plant in solution having different levels of osmoticum. Performance of seedling in response to osmotic stress is judge with respect to their moisture stress resistance ability.

Leaf Water Retention Method

Leaves are excised and weighed immediately and allowed to dehydrate under standard environment. Relative moisture loss after a period is determined. Less reduction in weight has been associated with drought resistance ability in some of the crops.

Evaluation based on RWC, Water Potential and, its Components and Osmotic Adjustment

Measurement of RWC and water saturation deficit (WSD) is an initial step to quantify plant water status. Under moisture stress, stress resistant plants tend to maintain higher RWC. It is observed that primary effect of mild moisture stress on plants is regulated by change in turgor pressure, and movement of water in soil-plant-atmospheric continuum (SPAC) is in response to water potential gradient, therefore, it is suggested that the measurement of leaf water potential, osmotic potential and turgor pressure would give a better understanding of plant water status (Srivastava and Tyagi, 1999). Under moisture stress, resistant genotypes maintain higher water potential, and higher reduction in osmotic potential due to active accumulation of solutes; which is termed as osmotic adjustment and leads to turgor maintenance and water absorption capability in plants (Singh *et al.*, 1990; Morgan, 1983).

Evaluation Based on Cell Membrane Injury

Cell membrane is the primary target of abiotic stresses. Nevertheless, a stable cell membrane that remain functional during water stress appear central to adaptation to high temperature and tolerance to drought (Singh *et al.*, 1992). A single measurement of cell membrane injury at 25 day after sowing have been found to be related with drought to tolerance in wheat (Singh *et al.*, 1992).

Biochemical Markers

The effect of stress on plants is reflected with derangement in metabolic changes. During stress there is change in the concentrations of various cellular constituents. Cellular concentrations of some of the compounds have been used as marker to identify resistant genotypes.

Plant hormones, generally cytokinins and ABA contents have also been associated with drought resistance in plants. Soil drying results in increased synthesis of ABA in roots, which moves to shoot and inhibits stomatal opening and leaf growth (Davis *et al.*, 1994). The concentration of ABA in xylem stream has been found to be a sensitive indicator of root zone soil moisture status.

Evaluation Based on Transpirational and Photosynthetic Parameters

Parameters like transpiration rate, leaf conductance, cuticular conductance, leaf temperature, leaf to air temperature difference are widely used as screening criteria (Srivastava and Chaturvedi, 1989). Introduction of IRGA, steady state parameter or diffusion parameter has further made these studies easier and rapid.

The IRGA is effectively used to screen for high photosynthetic rates. Introduction of fluorescence meter has also made field screening of resistant genotype on the basis of photosynthetic properties.

Carbon Isotopic Discrimination Technique

Carbon occurs naturally as two stable isotopes ^{12}C and ^{13}C. Most carbon is ^{12}C (98.8 per cent). ^{12}C diffuses faster than ^{13}C. The RUBISCO fixes ^{12}C faster than ^{13}C. The link between $^{13}C/^{12}C$ discrimination (d^{13}) and efficiency of water use is plants (WUE) is via the gas exchange characteristics of leaves. As RUBISCO discriminates against ^{13}C, ^{13}C is concentrated relative to ^{12}C in the intercellular spaces resulting in RUBISCO fixing more ^{13}C relative to ^{12}C and d^{13} decreases. This is reflected in carbon isotope ratio in C_3 plants. In C_3 plants, a lower ^{13}C discrimination is associated with higher to water use efficiency.

On the Basis of Yield Stability Index

There are various ways to calculate stability index and susceptibility index. The widely accepted one is by Ficher and Maurer, 1978 as:

$$Yd = Yp\ (1 - SD), \text{ and}$$

$$D = 1 - (Xd = Xp),$$

Where, Yd = stress yield, Yp = potential yield under non-stress, Xp and Xd are average yields of all genotypes under and stress condition respectively, D = intensity of drought and S = drought susceptibility index.

For stability analysis genotypes are to be grown under normal (non-stress) and various levels of stress, which may be created by various methods *viz.*, sprinkler irrigation, multilocation trials or in lysemeters (Srivastava and Rami, 1998).

Conclusion

There are a number of screening techniques for stress resistance in plants. No single technique can be recommended as an ideal technique to screen all crop plants under various conditions. Nevertheless, depending upon species under investigation, growth stage, location and facilities available a technique or combination of techniques may be adopted to screen crops and their genotypes for a particular stress.

References

Buchanan, B.B.; Gruissem, W. and Jones, R.L. (2000). Biochemistry and molecular biology of plants. *Amm. Soc. Plant Physiologists.* Rockville, Maryland, USA.

Chaturvedi, G.S. (Personal communication) Department of Crop Physiology. ND University of Agric. and Tech. Faizabad.

Davies, W.J., Tardieu, F. and Trejo, C.L. (1994). How do chemical signals work in plants that grow in dry soil. *Plant Physiol.* 104 : 303-314.

Fiseher, R.A. and Maurer, R. (1978). Drought resistance in strong wheat cultivars. 1. Grain yield responses. *Aust. J. Agric. Res.* 29: 897-912.

Gates, D.M.C. (1968). Transpiration and leaf temperature. *Ann. Rev. Plant Physiol.* 19: 211-238.

Idso, S.B., Reginato, R.J., Reicosky, D.C. and Hotfield, J.L. (1981). Determining soil induced plant water potential depression in alfalfa by means of infrared thermometry. *Agronomy J.* 73: 826-830.

Levitt, J. (1972). Response of plants to environmental stresses. In : *Physiological Ecology* Academic Press, N.Y.

Morgan, J. M. (1983). Osmoregulation as a selection criteria for drought tolerance in wheat. *Aust. J. Agric. Res.* 34: 607-614.

O' Tool, J.C.O., Cruz, R.T. and Singh, S.N. (1979). Leaf rolling and transpiration. *Plant Sci., Letter* 16: 111.

Srivastava, J.P. and Chaturvedi, S.N. (1989). Influence of water deficit on transpirational and water relation parameters in wheat. *Ann. Zone.* 28 :257-266.

Srivastava, J.P. and Kumar, A. (1994). Current perspective in water loss from plants. In : M. Pessarakli, ed., *Handbook of Plant and Crop Physiology Marcel Dekker*, N.Y. pp. 45-60.

Srivastava, J.P. and Ram, K.C. (1998). Analysis of yield stability in seasame (*Sesasumum indicum* L.). *Indian J. Plant Physiol.* 4 (MS): 256-259.

Srivastava, J.P. and Tyagi, D.N. (1999). Advances in water relation and moisture stress stueies of plants. In : M. Pessarakli, ed., *Handbook of Plant and Crop Physiology Marcel Dekker*, N.Y. pp. 1029-1039.

Singh, M., Srivastava, J.P. and Kumar, A. (1990). Effect of water stress on water potential components in wheat genotypes. *Indian J. Plant Physiol.* 28: 312-317.

Singh, M., Srivastava, J.P. and Kumar, A. (1992). Cell membrane stability in relation to drought tolerance in wheat genotypes. *J. Agron. Crop Sci.* 168: 186-190.

Chapter 10

Water in Soil-Plant-Atmosphere Continuum in Relation to Moisture Stress

J.P. Srivastava

Department of Plant Physiology,
Institute of Agricultural Sciences, Varanasi – 221 005, U.P.

About 70 per cent of earth surface is covered by ocean, and it contains about 97 per cent of total water on earth. The atmosphere contains about 0.00001 of the planets total water (Hart, 1986). Turnover of atmospheric water occurs 40 times in a year *i.e.*, roughly every 9 days.

Water is essential to the life on earth. Human body constitutes nearly 85 per cent water, while vegetative plant materials may contain closer to 90 per cent moisture. A typical crop or grassland may transpire 500 kg water per kg dry matter production (Volkmar and Woodbury, 1994). Of all water absorbed by plants, only 5 per cent or even less is utilized; the remaining 95 per cent is lost (Srivastava and Kumar, 1994).

Water being hydrophilic is a good solvent, constituent of protoplasm and other cell constituents, and participates in a number of biochemical reactions. Availability of water to plants, in terms of quantity, distribution and quality are attaining greater significance in crop production. Inadequate supply of moisture leads to moisture stress in plants. Crop water deficit or moisture stress is the result of complex interactions among factors in the rhizosphere, the plant and the atmosphere. Plant water status is determined by the balance between absorption from soil and loss through transpiration to atmosphere. To understand the mechanism that how water is absorbed, transported from root to shoot, and finally transpired, it is essential to quantify water in soil-plant-atmosphere continuum (SPAC).

Basics of Plant Water Relation

A number of ways have been used to quantify plant water status, *viz.* absolute water content, relative water content, water saturation deficit, and so forth. But later on it was suggested that chemical potential be used as the basis for the property of water in plant as well as in soil. However, if chemical potential of water is divided by specific volume of water which is 1 $cm^3 g^{-1}$, the energy unit is expressed in the unit of pressure.

Water potential (y_w) of pure water is zero. When a solute is dissolved in water, the water potential of the system becomes negative, as the presence of solutes decreases the free energy of water. Substances in a liquid or gas phase move in response to differences in chemical potential.

In a plant water potential is mainly governed by osmotic potential (ys) and pressure potential (yp) or turgor pressure. Osmotic potential of plant is always negative. Pressure potential of a turgid cell is positive. In living system there is another component of water potential and that is matric potential (ym) it is negative and very low in magnitude.

There are a number of ways to quantify various components of plant water potential (Srivastava and Tyagi, 1999).

In SPAC water moves from higher water potential to lower water potential, *i.e.* leaf and shoot water must be lower than the root and soil y. Transpiration by reducing leaf water content and y, gives rise to the gradient for uptake. Another factor which regulates the flow of water is the resistance created by water transport pathway. In plants major water absorptive organ is root, while stomata regulate transpirational losses. If loss exceeds transpiration water deficit occurs in plants.

Various morpho-physiological and biochemical parameters respond differently to reduced plant water status (Hsiao *et al.*, 1976). Pressure potential is given as the major controlling factor of plant responses under stress condition. Since cell enlargement is very sensitive to moisture stress, therefore, exponential growth phase of crop is very sensitive phase to moisture stress, as it leads to reduced canopy development. Crop yield is the result of net carbon dioxide assimilation and the partitioning of assimilates to the economic part. Both of these parameters are adversely affected under stress condition.

Supply of food grain for ever increasing population is possible mainly by improving productivity from stress constraint regions. One approach to improve productivity from such regions is by growing stress resistance crop and their genotypes. To achieve these objectives quantification of water in plant water atmosphere continuum and the studies on impact of stress on plants processes shall be of immense importance.

References

Hart, J.C. (1986). Consider a spherical cow. A course in environmental problem solving. Kaufmann. Los Altos, CA. pp. 238.

Hisao, T.C.; Acevedo, E.; Fereres, E. and Henderson, D.W. (1976). Phil. Trans. R. Soc. London Ser. B. 274:479.

Srivastava, J.P. and Tyagi, D.N. (1999). Advances in water relation and moisture stress stueies of plants. In : M. Pessarakli, ed., *Handbook of Plant and Crop Physiology Marcel Dekker,* N.Y. pp. 1029-1039.

Srivastava, J.P. and Kumar, A. (1994). Current perspective in water loss from plants. In : M. Pessarakli, ed., *Handbook of Plant and Crop Physiology Marcel Dekker,* N.Y. pp. 45-60.

Volkmar, K.M. and Woodburg, W. (1994). Plant water relationship. Handbook of Plant and Crop Physiology. (M Pessarkali ed.) pp. 23-43. Marcel Deckker Inc. USA.

Chapter 11

Plan for Starting a Dairy Unit

U.K. Pandey
Prof. (Livestock Economics)
Dept. of Veterinary and Animal Husbandry
CCS HAU, Hisar

Assumptions

1. Land for the construction of shed is available. Moreover, bank loan will be taken for the construction of shed and purchase of animals. A part of working capital will also be borrowed from the bank to start with.
2. The animals purchased will be freshly calved in 2nd and/or 3rd lactation having a female calf at its heel. Moreover, the animals will be purchased in phases over a period of six months.
3. In mixed herd, the buffaloes to cows ratio should be 60:40 for better viability.
4. Replacement of milch animals is more economical by raising heifers on the farm rather than purchasing freshly calved adult milch animals.
5. The amount of loan borrowed will be either total initial capital investment or 80 per cent of the total, whichever is applicable.

6. Loan repayment will be made in 60 instalments with the amortized repayment plan. However, the period of moratorium needed for setting up of infrastructure may be negotiable.
7. There shall be adequate and timely availability of veterinary aid and breeding facilities, green/dry fodders and concentrates, etc.

Financial Details of the Project

Option 1 (5 Murrah Buffaloes with their Followers)

A. Initial Capital Investment	Amount (in Rs.)
1. **Cost of animals:** 5 lactating murrah buffaloes @ Rs. 20,000 each.	1,00,000.0
2. **Cost of shed:** Cost of shed for 5 buffaloes, 5 calves, including store-cum-office room covered area (4 sqm./animal @ Rs.800/sqm. and open area (8 sqm./animal @ Rs.200/sqm.	50,000.0
3. **Cost of equipments:** Such as milk cans, chains, milking buckets, weighing balance, chaff cutter, belcha, rickshaw/bullock cart etc.	60,000.0
4. **Initial variable expenses**	30,000.0
Total	**2,40,000.0**

Note: The initial total capital investment exceeds Rs. 2.00 lac, an amount which is generally available under Prime Minister Rozgar Yojana. Hence, the maximum limit for bank loan is Rs. 2.00 lakh while remaining amount the entrepreneur will add as margin money.

B. Financial Pattern

Bank loan	= Rs. 2.00 lac
Margin money	= Rs. 0.40 lac
Total	= Rs. 2.40 lac

C. Expenses

(a) Fixed Cost	***Amount (in Rs.)***
1. Depreciation on shed @ 5 per cent/annum	2500.0
2. Depreciation on machinery and equipments @ 10 per cent/annum	6000.0
3. Interest on bank loan @ 10 per cent/annum	20000.0
4. Insurance premium of milch animals @ 4 per cent/annum	4000.0
Total fixed cost	**32500.0**

(b) Variable Cost	
1. **Cost of green fodder:** 30 kg/buffalo/day *i.e.* 547.5 qtls. @ Rs. 65/qtl.	35,587.5
2. **Cost of dry fodder:** 4 kg/buffalo/day *i.e.* 73 qtls. @ Rs.175/qtl.	12,775.0
3. **Cost of concentrate:** 3 kg/buffalo/day *i.e.* 54.75 qtls. @ Rs.700/qtl.	38,325.0
4. **Labour expense:** At the rate of Rs.100/day/labour for 300 days	30,000.0
5. **Veterinary care:** At the rate of Rs.400/head/annum	2,000.0
6. Misc. expenses	8,000.0
Total variable cost	**1,26,687.5**
Total cost (a+b)	**1,59,187.5**

(c) Income	
1. **Sale of milk:** 8 kg/buffalo/day for 280 days *i.e.* 11,200 kg @ Rs. 16/kg	1,79,200.0
2. **Sale of manure:** At the rate of Rs.1000/head/annum	7,000.0
3. **Appreciation value of heifers:** At the rate of 2500/head/annum	12,500.0
Total returns	**1,98,700.0**

Net income/annum (C – (a+b)	**39,512.5**
Net income/month	**3292.7 or say Rs. 3293.0**

Option 2 (5 Crossbred Cows with Followers)

A. Initial Capital Investment	**Amount (in Rs.)**
1. **Cost of milch animals:** 5 lactating crossbreds @ Rs. 20,000 each	1,00,000.0
2. **Cost of shed:** Cost of shed for 5 crossbreds and youngstock, store-cum-office covered area (4 sqm./animal @ Rs.800/sqm.) and open area (8 sqm./animal @ Rs.200/sqm)	50,000.0
3. **Cost of equipments:** Such as milk cans, chains, milking buckets, weighing balance, chaff cutter, belcha, rickshaw/bullock cart etc.	60,000.0
4. **Initial variable expenses**	30,000.0
Total initial capital investment	**2,40,000.0**

Note: The initial total capital investment exceeds Rs. 2.00 lac an amount which is generally available under Prime Minister Rozgar Yojana. Hence, the maximum limit for bank loan is Rs. 2.00 lac while remaining amount the entrepreneur will add as margin money.

B. Financial Plan

Bank Loan	= Rs. 2,00,000
Margin money	= Rs. 40,000
Total	= Rs. 2,40,000

C. Expenses

(a) Fixed Cost	***Amount (in Rs.)***
1. Depreciation on shed @ 5 per cent/annum	2,500.0
2. Depreciation on machinery and equipments @ 10 per cent/annum	6,000.0

3. Interest on bank loan @ 10 per cent/annum	20,000.0
4. Insurance premium of milch animals @ 4 per cent/annum	4,000
Total fixed cost	**32,500**

(b) Variable expenses

1. **Cost of green fodder:** 30 kg/crossbred/day *i.e.* 547.5 qtls. @ Rs.65/qtl.	35,587.50
2. **Cost of dry fodder:** 4 kg/crossbred/day *i.e.* 73 qtls. @ Rs.175/qtl.	12,775.0
3. **Cost of concentrates:** 3 kg/crossbred/day *i.e.* 54.75 qtls. @ Rs.700/qtl.	38,325.0
4. **Labour expenses:** At the rate of Rs.100/day/labour for 300 days	30,000.0
5. **Veterinary care:** At the rate of Rs. 400/head/annum	2,000.0
6. Misc. expenses	8,000.0
Total variable cost	**1,24,687.5**
Total cost (a+b)	**1,59,187.5**

(c) Income

1. **Sale of milk:** 10 kg/cow/day for 300 days *i.e.* 15,000 @ Rs. 12/kg	1,80,000.0
2. **Sale of manure:** At the rate of Rs. 1000/head/annum	7,000.0
3. **Appreciation value of heifers:** At the rate of Rs. 2500/head	12,500.0
Total returns	**1,99,500.0**
Net income/annum (C – (a+b)	40,312.5
Net income/month	**3359.4** or **3359.0**

Option 3 (Mixed Herd of 6 Murrah Buffaloes and 4 Crossbred Cows)

A. Initial Capital Investment	**Amount (in Rs.)**
1. **Cost of milch animals:** @ Rs. 20,000 each	2,00,000
2. **Cost cattle shed :** Shed for 10 milch animals, 10 youngstocks, office room, store room etc. covered area (4 sqm./animal @ Rs. 800/sqm.) and open area (8sqm/animal @ Rs.400/sqm.)	1,00,000.0
3. **Cost of equipments:** chaff cutter, bullock cart/ rickshaw, belcha, chains, etc.	90,000.0
4. Initial variable expenses	35,000.0
Total initial capital investment	**4,25,000.0**

Note: The initial capital investment exceeds Rs. 2.00 lac of PMRY. Moreover, efforts will be made to approach the lead bank/ NABARD/Kshetriya Gramin Bank to get 80 per cent amount needed.

B. Financial Plan

Bank Loan:	= Rs. 3,40,000
Margin money	= Rs. 85,000
Total	= Rs. 4,25,000

C. Expenses

(a) Fixed Cost	***Amount (in Rs.)***
1. Depreciation on shed @ 50 per cent/annum	5,000.0
2. Depreciation on equipments @ 10 per cent/annum	9,000.0
3. Interest on bank loan @ 10 per cent/annum	34,000.0
4. Insurance premium @ 4 per cent/annum	8,000.0
Total	**56,000.0**

(b) Variable expenses

1. **Cost of green fodder:** 30 kg/animal/day *i.e.* 1095 qtls @ Rs.65/qtl.	71,175.0

2. **Cost of dry fodder:** 4 kg/day/animal *i.e.* 146 qtls. @ Rs.175/qtl.	25,550.0
3. **Concentrate:** 3 kg/day/animal *i.e.* 109.5 qtl.s @ Rs.700/qtl.	76,650.0
4. **Labour expenses:** One labour for a year at the rate of Rs.100/day per labour	36,500.0
5. **Veterinary Care:** At the rate of Rs.400/head/annum	4,000.0
6. Misc. expenses	16,125.0
Total variable cost	**2,30,000.0**
Total cost (a+b)	**2,86,000.0**

(c) Income

1. **Sale of cow milk:** 12,000 litres (10 kg/day/cow for 300 days) @ Rs. 12/kg	1,44,000.0
2. **Sale of buffalo milk:** 13440 litres (8 kg/day/buffalo for 280 days) @ Rs. 16/kg	2,15,040.0
3. **Sale of manure:** At the rate of Rs. 1000/head/annum	15,000
4. **Appreciation value of heifers:** At the rate of Rs. 2500/head/per annum	25,000
Total income	**3,99,000**
Net income/annum (D-C)	1,30,000.00
Net income/month	**9416.6 or 9417.0**

Chapter 12

Marketing of Buffaloes in Haryana

U.K. Pandey
Prof. (Livestock Economics)
Department of Veterinary and Animal Husbandry
CCS HAU, Hisar

Livestock development starts at the marketing and as it provides the necessary incentive to the producers to adopt the scientific methods of animal husbandry. The marketing of buffaloes, which constitutes a big business to the farmers and animal dealers, is mostly done through the seasonal cattle fairs of unregulated markets and that too under non-systematic lines. The most prevalent channels of marketing buffaloes in Haryana include:

1. Producer → consumer/final buyer
2. Producer → trader → consumer/final buyer.
3. Producer → village agent → trader → consumer/final buyer and
4. Producer → local trader → trader in consuming area → consumer/final buyer.

The market price of an individual lactating and/or dry pregnant buffalo at a point of time and place depends upon factors such as the basic economic factors, those associated with the individual animal and miscellaneous factors like locality, season, etc.

Furthermore, both quantitative and qualitative factors have a bearing in determining the market prices of lactating/dry pregnant buffaloes.

The quantitative factors of lactating Murrah buffaloes are milk yield, stage and order of lactations. However, the qualitative factors include breed (non descript or Murrah), tail length (short or long), tail colour (mixed (black and white) or black), horns (straight or curly), temperament (erratic or cool and calm), size of teats (abnormal or normal), seasons of sale/purchase (winter or summer), attachment and placement of udder (pendulous udder/broken udder with uneven quarters or high rear attachment and well forward placement of udder) and general appearance (bad or good).

A freshly calved Murrah buffalo of second lactation and yielding about 12 litres of milk per day has the potential to fetch the maximum market price. Likewise, a lactating Murrah buffalo with long tail and carrying a good switch, mixed tail colour, curly horns, cool and calm temperament, uniform and normal size of teats, high rear attachment and well forward placement of udder as well as having a good general appearance if sold in the summer season has the potential to fetch higher market price.

It has been observed that in every hundred rupee market price of lactating Murrah buffaloes, the share of milk yield, order and stage of lactations was about 47.3 and 7 rupees, respectively. Similarly, beyond peak milk yield as well as optimum order and stage of lactations the share of these characters in every hundred rupee of declining market price of lactating Murrah buffaloes was about 15, 3 and 11 rupees, respectively. Thus among the quantitative characters, the milk yield made the highest contribution to the market price of lactating Murrah buffaloes followed by the stage and order of lactations. Among the qualitative characters, the general appearance had the highest score in the market price of lactating Murrah buffaloes, followed by breed, size of teats, temperament of animal as well as the attachment and placement of udder. These qualitative characters together accounted for a share of about 9 per cent in the market price of lactating Murrah buffaloes. However, the share of remaining qualitative characters (tail length, tail colour, season of sale/purchase and horns of the animal) in the market price of lactating Murrah buffaloes was only about 4 per cent (Table 12.1).

Table 12.1: Relative Contribution of Various Factors Influencing the Market Price of Lactating and Dry Pregnant Murrah Buffaloes

Sl.No.	Lactating Buffaloes		Dry Pregnant Buffaloes	
	Factors/Characters	Percentage Contribution	Factors/Characters	Percentage Contribution
1.	Milk yield (X_1)	46.56	Stage of pregnancy (X_1)	48.04
2.	X^2_1	15.39	X^2_1	15.56
3.	Order of lactation (X_2)	3.36	Age (X_2)	15.24
4.	X^2_2	3.17	X^2_2	11.76
5.	Stage of lactation (X_3)	6.78	Non-descript breed (X_3)	1.83
6.	X^2_3	11.37	Short-tail length (X_4)	1.45
7.	Non-descript breed (X_4)	2.04	Mixed-tail colour (X_5)	0.48
8.	Short tail length (X_5)	1.14	Erratic temperament (X_6)	0.78
9.	Mixed tail colour (X_6)	1.04	Abnormal size of teats (X_7)	0.52
10.	Straight horns (X_7)	0.90	Pendulous/broken udder with uneven quarters (X_8)	1.37
11.	Erratic temperament (X_8)	1.34		
12.	Abnormal size of teats (X_9)	1.53	Bad general appearance (X_9)	2.87
13.	Winter season (X_{10})	0.93		
14.	Pendulous udder/broken udder with uneven quarters (X_{11})			
15.	Bad general appearance (X_{12})	3.14		

Source: 1. P.K. Mondal and U.K. Pandey, "Factors Influencing the market price of lactating Murrah buffaloes in Haryana", *Indian J. of Agril. Econ.* Vol. 48, No.4, Oct-Dec., 1993, P. 692.

2. P.K. Mondal and U.K. Pandey, "Factors influencing the market price of dry pregnant Murrah buffaloes in Haryana", *Indian J. Dairy Sci.* 48(1), Jan., 1996, P.12-17.

Age and stage of pregnancy are the quantitative factors of dry pregnant Murrah buffaloes. Moreover, the qualitative characters include breed (Non-descript or Murrah), tail length (short or long), tail colour (mixed (black + white) or black), temperament (erratic or cool and calm), size of teats (abnormal or normal), attachment and placement of udder (pendulous/broken udder with uneven quarters or high rear attachment and well forward placement of udder) and general appearance (bad or good).

A dry pregnant buffalo of advance stage (between 8th and 9th month) and of 5th or 6th year of age commanded the higher market price. There existed a price discount for the non-descript breed over Murrah, short tailed over long tailed ones, the erratic temperament of an animal over the cool and calm, the abnormal size of teats over the normal, pendulous (swinging) udder/broken udder with uneven quarters over high rear attachment and well forward placement of udder and bad general appearance over the good general appearance. However, the mixed tail colour had premiums over black.

The respective shares of stage of pregnancy and age pertained to about 48 and 15 rupees in every hundred rupees market price of dry pregnant buffaloes. Similarly, beyond optimum age and below optimum stage of pregnancy, the respective shares of these characters in every 100 rupee declining market price of dry pregnant buffaloes pertained to about 12 and 15 rupee. Amongst qualitative characters the general appearance had the highest score into the market price of dry-pregnant buffaloes followed by breed, tail length, attachment and placement of udder as well as the temperament of animal. These qualitative characters together accounted for about 8 per cent share into the market price of dry pregnant buffaloes. However, the remaining qualitative characters such as tail colour and size of teats had only about 1 per cent share with the market price of dry pregnant buffaloes.

The net share of producer in the consumer's rupee price had been worked out about 88 per cent (Table 12.2). Indeed, it had improved by about 8 per cent during the last decade or so. Studies have also shown that the retail prices of buffaloes were minimum in October and maximum in July. Furthermore, the retail prices of buffaloes increased by about 8 per cent in July, declined by about 4 per cent in October, and were significantly different in these months as compared to that in other months. The retail prices of buffaloes showed highest variability in February and lowest in December. In general, the retail/prices of buffaloes significantly increased both between years and months. In absolute terms, the retail prices of buffaloes though increased, yet in real terms (at constant prices of 1980) declined thereby indicating that the buffalo rearers were not losers over time. As compared to winter and rainy seasons, the retail prices of buffaloes had increased faster during summer.

Table 12.2: Price Spread for Marketing of Buffaloes in the Markets of Haryana

Item of Cost	*Amount (In rupees)*	*Percentage of the Consumer's Price*
1. Net price of producer/seller	19,355.6	87.98
2. Costs incurred by the sellers		
(i) Preparation of buffalo before sale (feed and labour)	486.2	2.21
(ii) Transportation charges	369.6	1.68
(iii) Market charges, *i.e.*, Registration fee @ Rs. 6 per animal	6.0	0.03
(iv) Brokerage charges @ 1 per cent of the sale price	202.4	0.92
3. Consumer's purchase price	20,420.4	92.82
4. Costs incurred by consumer/buyer		
(i) Buffalo feed and labour costs after sale	145.2	0.66
(ii) Transportation charges	486.2	2.21
(iii) Sale registration charges @ 4 per cent of the sale price	816.2	3.71
(iv) Other taxes *i.e.* sales tax	132.0	0.60
5. Consumer's price	22.000.0	100.00

Source: P.K. Mondal and U.K. Pandey, "Price spread in Marketing of Milch Buffaloes in Haryana", *Agricultural Marketing Journal*, 36(3), Oct-Dec., 1993, p. 28.

Chapter 13

Use of Oxytocin in Milch Animals: A Boon or Bane

V.N. Bhargava

Prof. (Retd.) of Animal Husbandry and Dairying, B.H.U., Varanasi

Director (Operations), K.S. Singhal Dairies Pvt. Ltd., Agra

Wrong Signal

Several reports which appeared in print and electronic media have given an impression to the general public that use of Oxytocin injection for milking cows and buffaloes can have harmful effects on the health of both, the milk consumers and the producing animals. It is being disseminated that its use not only causes, pain, cancer and infertility in cows and buffaloes but consumption of milk from Oxytocin treated animals might result in health hazard including abortion in pregnant woman, early maturity in girls and impaired bone and brain formation in growing kids (Anon, 2000, 2005). Use of the this hormone for milking cows and buffaloes has been declared a legal offence under Prevention of Cruelty, in animals Act by the U.P. State Administration (Anon, 2000). These reports however, have no scientific basis (Singh, 2003).

Oxytocin and Milk Ejection

Oxytocin is an indispensable life saving drug used both in medical and Veterinary Practice "Oxytocic" in Greek means rapid birth. Ott and Scott (1911) reported a powerful galactogic effect on the mammary gland by neurohypophyseal extracts. Oxytocin is a polypeptide hormone which was first synthetically produced by Du Vigheud *et al.* (1959) from the posterior pituitary of Ox. Since then, if has been studied in great details and there are numerous reports about its useful functions in the body.

Oxytocin is released from the posterior lobe of pituitary gland as a result of nervous stimulation of teats in lactating mother or cow due to suckling by infant or calf, and plays a significant role in the "let down" of milk. To understand and appreciate the significance of this statement let us recall the events taking place during hand milking of cows or buffaloes.

Milk remains stored in two parts in the udder. The part stored in teats and mammary gland cisterns is called the "Cistern Milk". The other part stored in the milk producing alveoli of the mammary gland is called as "alveolar milk". In hand milking, if no washing and no preliminary massage of udder and teats are done, and the act of milking is started, then few streams of "cistern milk" will come out from each of the teats. This flow of milk will then stop, howsoever, skilfully fingers are moved over the teats with increased pressure. After a pause, intramammary pressure increases, udder and teats become turgid due to release of "alveolar milk" into the udder cistern and teats. This phenomenon is known as "milk let down" or "milk ejection". The flow of milk gets resumed and continues till milking is complete. In other words, the cistern milk can be removed by contraction of the smooth muscles surrounding the alueoli of the mammary glands, under the influence of hormone Oxytocin. Milking stimulus due to suckling by calf causes a neuro-endocrine reflex which is transmitted from the nerve endings of the teats, through the ascending nerves spinal cord and mid brain inducing the release of Oxytocin from the pituitary gland to the blood stream. This Oxytocin thus released in the blood, binds to the receptors of smooth muscles of udder surrounding the alevoli resulting in the their contraction and ejection of their contents that is preformed milk. Milk ejection reflex involves two areas – one comprising of nervous pathway and the other of release of Oxytocin.

By adopting better milking management practices about 90 per cent of the available milk in the udder can easily be removed under the influence of endogenous Oxytocin released due to milking stimulus. The quantity of milk left in the udder after normal milking can be removed by administration of exogenous Oxytocin and is known as residual milk.

Besides let down of milk, Oxytocin also possesses functions like luteolytic action. Vaginal distension in cows causes an increased release of Oxytocin, which is the final singal to cause a sudden release of Oxytocin needed for complete expulsion of the foctus during normal parturition. It is also released during coitus in both sexes in humans and may affect sperm transport through the male tract and may have role in maintenance of relationships (Turner, 1999). It is also reported to be responsible for developing trust between mother and the newborn (Anon (a) 2005). But of all the function, milk ejection reflex is the best established function of Oxytocin (Singh and Dang, 2002)

Concentration of Oxytocin in Relation to Functions

Plasma Oxytocin concentration which is essential for causing the milk ejection is very low and is degraded rapidly by enzyme Oxytocinase giving the plasma Oxytocin a half-life period of 2-3 minutes (Wakerley *et al.*, 1994). Sight of calf and milker, sound of milking pails to the cow are effective in releasing Oxytocin in blood stream, which further increases when udder is washed with luckewarm water. However, sucking is more effective than any of the factors associated with milking. Stimulation of genital tract caused greater Oxytocin release than mammary gland stimulation.

Recommended dose of Oxytocin for letting down of milk is as low as 1-3 internation units (IU). It needs to be appreciated that it is naturally occuring hormone which is relatively unstable and is inactivated in 2-3 minutes following endogenous release or exogenous injection. It is produced in all mammals for letting down milk during each milking and for induction of labour at term to facilitate parturition. However, far its action on uterine muscles much larger amounts (8-10 times) are required (Singh, 2003). Therefore the propaganda that milk from Oxytocin treated animals can cause uterine contractions and abortions in human is baseless.

The so-called ill effects of Oxytocin on length of oestrus cycle, causing cancer and pain etc., are also baseless and unscientific. Studies carried out at Cornell University, New York have demonstrated that cows treated with exogenous Oxytocin even at dose of 20 IU for each milking, twice daily for 305 days had no apparent detrimental effect on their health particularly on the length of oestrus cycle. Further, works of Italian scientists (1997) and chinese workers (1998) also indicated that Oxytocin can be advantageously used for treating infertility in cows (c.f. Singh 2003).

Therapeutic Use

Besides its use in treating "holding up" of milk, Oxytocin as a drug has also been recognised for treating uterine inertia during parturition and prevention of hemorrhage after birth particularly in Oxytocin-deficient females. It is also used as an adjunct to antibiotic therapy for treating mastitis, for removal of residual infected milk (Knight *et al.*, 2000). Because of its therapeutic values, it is included in list of essential drugs and in the latest edition of Indian Pharmacopoeia (1996, cf. Singh 2003).

Effect on Yield and Composition of Milk

It has been demonstrated that suckling stimulus by calf causes higher yield of milk, than by injection of Oxytocin given evening and morning but provided no calf stimulus (Bhular *et al.*, 1991). Contrary to this Nostrand *et al.* (1991) reported that cows administered 20 IU Oxytocin at each milking produced 849 Kg more milk in the lactation than the control group.

So far the nutritive value of milk from Oxytocin treated cattle is concerned, researches at Cornell University, U.S.A. and N.D.R.I. Karnal have clearly showed that daily administration of Oxytocin over a long period, in appropriate doses had no effect on the composition of milk (Ludri, 1980).

Further, this peptide hormone following entry into blood stream produces the desired effect within a minute by acting on the target smooth muscles having receptors for its action and get rapidly metabolized leading to inactive substances. Till date, there is not a single report, which confirms the presence of this hormone in milk (Singh, 2003). The question of harmful effects to consumers of milk obtained from Oxytocin treated animals therefore does not arise at all.

In view of the aforesaid, we must realise that when endogenous Oxytocin in response to natural stimuli is not released due to death of calf, fright, noise or completely new surroundings, the use of exogenous Oxytocin has to be adopted. Similarly, in the absence of adequate concentration of endogenous Oxytocin, if uterine contractions for normal delivery are impaired the use of exogenous Oxytocin is essential, most rational and harmless.

Thus Oxytocin is a boon, but to prevent its misuse, it should be used under strict supervision and advice of a veterinarian. Researches on quantity of Oxytocin which may appear, if any in milk of oxytocin administered cows buffaloes are needed to confirm its harmless beneficial effects.

References

1. Anon (2000) "Jyada Doodh Pane Ke Liye Maare Ja Rahe Hain Gaay Bhains Ke Bachhe" in Dainik Jagran, Agra (ed) dated 27th April 2000 page 3.
2. Anon (2005) "Oxytocin : Doodh Me Dheema Jahar" in Dainik Jagran, Agra (ed) dated 30th June 2005. page 4.
3. Anon a (2005) "Test reveal 'trust' in a bottle". In CNN.Com-(Internet) Int. Ed. dated 1st Jun, 2005.
4. Bhular M.S., Gurcha, D.S, and (1991) In Third Buffalo Congress Warn, Bulgaria.
5. Du Vigheud, V., Rester, C. and Tripett, S (1953) The Sequence of amino acid in Oxytocin with a proposal for structure of Oxytocin. J. Biol. Chem. 205:949.
6. Kright, C.H., Filzpatrick, J.L., Lague, D.N. and Platt D.J. (2000) Efficacy of two non-antibiotic therapies, Oxytocin and tropical Liniment against bovine master Vet. Rec. 146:311-316.
7. Ludri R.S. (1980) Note on milk let down response and residual milk in buffaloes. Indian Journ. Amin. Sci. 50:891-893.
8. Nostrand S.D., Gatton, D.M. Erb., H.N. and Bauman, D.E. (1991) Effects of daily exogenous Oxytocin on lactation milk yield and composition. J. Dairy Sci. 74:2119-212.
9. OH, I.S., and Scott., J.C. (1911) The action of infundibulum upon the mammary secretion Proc. Soc. Exp. Bio. Med. 8:48-49.

10. Singh, D.K. (2003) Facts about Oxytocin: The life saving drug. Indian Dairyman 55(8) 29-32.

11. Singh, M. and Dang, A.K. (2002) Does Oxytocin affect milk production and reproduction of dairy animals. Indian Dairyman 54(4) 65-69.

12. Turner (1999) Hormone involved in reproduction may have role have role in maintenance of relationship. (Internet) CNN. Com.

Chapter 14

Alternative Therapies for Treatment of Dairy Animals

V.N. Bhargava
Prof. (Retd.) of Animal Husbandry and Dairying, B.H.U., Varanasi
Director (Operations), K.S. Singhal Dairies Pvt. Ltd., Agra

Introduction

In general, animal behaviour is no different from the one of homosapiens. This is reflected from the following assertion: "animals have souls and conscious awareness, They experience all the same emotions that human beings experience, think and feel. They experience sadness and grief. They can communicate their thoughts and feelings to us and beautiful ways of looking at ourselves, our lives and world in general."

Sharon Callahan

Undoubtedly, all animals should receive treatment like human beings. Nature has now compelled us to move in this direction. Pure fresh milk is the dream of most of the citizens but adulteration, presence of chemical additives and residues in it has become one big worry for the consumers. The rising quality consciousness among the consumers has created demand for organic milk and its products apart from its export potential. Production of organic milk can only be done by following rigidly defined standards and conditions for feeding, management of dairy animals. Use of allopathic medicines for veterinary treatment is mostly not permitted for organically maintained livestock. Alternative therapies using medicines from natural origin *viz.*, herbal, ayurvedic, homeopathic and other age old indigenous preparations and techniques should be made use of for prevention and cure of diseases in organic livestock husbandry.

Salient Alternative Therapies for Animals

In today's materialistic world simple remedies do not suit this luxurious generation, that longs for what is elaborate and costly. With this background, people often argue alternative therapies are no substitute for veterinary or allopathic medicines. Nothing would be wrong, no sentiment will be hurt, if we use what is most accessible and appropriate for a wide diversity of health problem. It will therefore, be better to use the term "complementary alternative therapies" for animals. Salient ones are given below:

Acupuncture

It is recognized as, "the examination and stimulation of specific points of the body of animals by use of acupuncture needles, injections, low level lasers, magnets and a variety of other techniques for diagnosis and treatment of various conditions in animals. It is used for variety of applications ranging from sedation to arthritis pain management, to treatment of deaf animals.

Magnetic Therapy

At first glance magnetic therapy might seem a bit strange or new age but by no means "quackish". Keep in mind that earth has a normal magnetic field and the cells of animal body also have magnetic field that allows for proper functioning. In NASA experiments, it was seen that rats in space that were not provided suitable magnetic field perished due to disrupted energy flow.

Magnets can be used as permanent magnets also called static magnets or pulsed electromagnetic field magnets (PEMF). Static magnets come in bars, beads or strips. PEMF use a pulsing current that flows through a wire coil to create magnetic field around the wire: the greater the amount of current flow, and greater the number of turns of the wire, the greater will be the magnetic field that forms. PEMF is used for treating nonunion fractures (that have failed to heal), hip disorders, osteoarthritis, in animals. Magnets appear to heal the body removing inflammation and restoring circulation. In fracture healing, the use of magnetic fields increases the adherence of calcium ions to the blood clot formed at the site of break. This allows proper formation of the callus that is necessary for fractures to heal properly.

Massage Therapy

Used for calming down and relief to the animal during stress conditions.

Healing Touch Therapy

It is an energy–medicine modality combining philosophies, techniques and applications to correct energy balance and healing in animals. This works on holistic concept to assist in treating injuries, illness, wound healing, behaviour problems and stress related issues. It works by releasing endorphins to establish a deeper relaxation and state of well being allowing the body to function at a grater level. By allowing muscles to relax, circulation is increased, sending more oxygen, nutrients and hormones sending more oxygen, nutrients and hormones into the body to support healing.

Crystal therapy and vibration therapy also work in a similar fashion.

Herbal and Ayurvedic Therapies

Plant based medicines which are used in the so-called herbal or ayurvedic therapy, are often the most accessible and appropriate for a wide diversity of ailments in animals. India has a long list of medicinal plants which have proven therapeutic values. Many companies in the country have started making ayurvedic or herbal preparations for treating common ailments of doing animals. Several common herbal medicines *e.g.* aloe, catechu, croton seed, ginger, arecanut, cinchona, aniseeds, peppermint etc. are already in vogue to treat animals. Recently Das *et al.* (2002) have complied information on plant materials available in India used in treating diarrhea, foot and mouth disease, mastitis, tympany and eye infection in dairy animals. Mukherjee and Dash (2001) have reviewed the efficacy of Indian herbal preparations for treating sub-clinical mastities in cows. Popularity of herbal medicines is increasing as these are cheaper easily available and above all do not affect milk quality.

Homeopathic Therapy

Homeopathy came to light in 1790 due the investigations of German physician Samuel Hahnemann. The basic principle of homeopathy is that like be treated by like. Diseases of domestic animals yield as promptly to homeopathic treatment as for man. It acts euqally efficiently in large and small animals (Day, 1995).

Why Use homeopathy?

No Side Effects

It humane and gentle system of medicine with no toxic side effect on animal – while correcting one disorder in the body it cannot create another. It cannot cause accidental abortion, nor can it damage an unborn foetus.

No Requirement for Laboratory Animal Experiment

It was developed for humans in human "guinea-pigs" and now animals reap the benefits.

No Residues in Tissues or Milk

This gives it a plus point when one considers the present market trend governed mostly by quality awareness of the community.

Cost

Homeopathic remedies are by and large, quite cheap in relation modern allopathic drugs.

No Real Dependence on Diagnosis

Homeopathy treats the animal itself rather than a named disease. As such it is not necessary to name the disease prior to treatment.

It is Non-Suppressive

It does not suppress the disease like modern drugs, it cures the body by removing the cause. For example, continuos antibiotic treatment of mastitis in cows, after a period may result in creating antibiotic resistance in the causal bacteria and inflection will continue. Such a problem in homeopathic treatment will not because homeopathic remedy will not be a threat to bacteria but because it acts by increasing the individual host animal's resistance to bacteria.

Efficacy

Remedies are efficient and required only in small doses, of few globules or few drops.

Compatible with Other Therapies

Homeopathic remedies can be given along with other therapies.

Nature of Homeopathic Remedies and Preparation

Homeopathic remedies are derived from all natural sources, animal or biological, mineral or plant and their preparation includes soluble extract of material in alcohol known as the mother tincture and then first dilution and them succession. The later essential for the procedure, and consists of vigorous shaking by special equipment either of two ways: (1) decimal system used in Germany and France and expressed by numerals/letters X and D and (Z) the centesimal system popular in Britain and U.S.A. and expressed by letter C. In centesimal system one drop is added to 99 drops of solvent and in the decimal (X) dilutions are obtained by adding one drop to 9 at each stage. By dilution the toxic effect of the material are eliminated and more energy is released. Greater the dilution, more powerful the medicine becomes.

Source of Drugs

1. *Mineral Kingdom*: Minerals, metals, chemicals organic and inorganic. *e.g.* sulphur, phosphors.
2. *Plants*: *Acorite napellus, Belladona, Arrica caleudala.*
3. *Animals*: *Lackesis, Naja tripudiam, Blatta orentalis.*
4. *Sarcodes*: Certain secretions from healthy animals are used *e.g.* adrenalin, thyrondinum from thyroid extract.
5. *Nosodes*: Pathological products from human or animal sources. *bacillinum.*
6. *Impoderabilia* : The immaterial, weightless belongs to this class *e.g.* vials containing alcohol is exposed to X-rays, or magnets.

Administration of Remedies

Usually remedy is allowed to dissolve on tongue of the animal. It can also be injected subcutaneouly. Former is better for absorption.

Selection of Remedy

For this you should have:

1. *Knowledge of disease*: That is sun total of all symptoms.

2. *Knowledge of medicine*: Symptoms produced by them in healthy individual by consulting materiamedica books and repertories.

Objections against Alternative Therapies

While organic farming demands the use of alternative therapies, there are opinions brewing against these. Unprover remedica can be dangerous. Should not believe that because they are natural, they are safe. This is not necessarily true. For example, the foxglove plant naturally produces digitalis glycosides that are toxic to heart at certain levels. A chemical is chemical whether produced in lab or produced synthetically. Natural toxins are just as harmful as synthetic ones. Infact natural product may be more toxic than the purified synthetic product. These are the argument targeted against the herbal medicines. In homeopathic remedies, preparation of dilution takes care of these.

Other alternative therapies *viz.*, aromatheraphy, light therapy, aeflexology, vibrational healing and nutrition therapy which are also being used for treating selected ailments of animals.

It is better to prefer what is locally available and effective and reliable.

References

1. Das, P. Das, S.K. Arya, H.P.S., Reddy, S.G., Mishra, A. (2002), "Inventory of indigenous technical knowledge in Agriculture"- Document-I, I.C.A.R., New Delhi, pp. 411.
2. Day C. (1995) "The homeopathic treatment of beef and doing cattle". Pnb. Beaconsfield publishers Ltd. Beaconsfield, Backs, England.
3. Mukherjee, R. and Dash, P.K. (2001) "Herbal Therapies hold the key in organic doing training: the case of mastitis". Indian Dairyman 53(12) pp. 40-42.

Chapter 15

Milk Quality Assurance by Selfimposition of Milk Quality Regulations

V.N. Bhargava

Prof. (Retd.) of Animal Husbandry and Dairying, B.H.U., Varanasi

Director (Operations), K.S. Singhal Dairies Pvt. Ltd., Agra

Present Scenario

Every now and then reports appear in the media about use of so called "Synthetic Milk" to adulterate market milk. Samples of milk taken from vendors, sweetmeat shops and private traders in Hyderabad city were found not to meet the minimum legal standard requirements (Rao *et al.,* 2002). Consumers feel worried about food safety and fear to consume Milk.

Traditionally, it has been the practice of the national governments to assume sole responsibility for regulation of food quality through various food laws, with aim to protect the health of consumers. Our food laws are good but enforcement is poor (Shourie, 1998). Primarily, because the existing regulatory systems are based on policing and end-product testing. Instances are not lacking when milk has been found adulterated with vegetable oils, urea, caustic soda, formalin and even palmolein. The laws are quite strong and stringent. Provisions for imposing penalty are adequate and the procedure for trial can be efficacious. But the enforcement is lacking. The trader and vendors feel comfortable and they do not fear. They in fact enjoy a "cosy" relationship with inspectors and also with testing laboratories. Under the conditions, prevailing, the poor and deprived sections of the population who necessarily depend on street milk vendors supplying loose unpacked milk remain the main victims.

Trade liberalization and globlisation present new challenges in respect of ensuring the quality and safety of dairy products for both domestic consumption and export. The Agreement on application of Sanitary and Phytosanitary Measures (SPS Agreement) under the WTO regime aims to introduce discipline in the international and domestic food trade and curb the use of unjustified protection measures. Consequently, all member countries are obliged to harmonize their dairy standards/guidelines with that of international rules as framed by the codex Alimentation Commission. The Indian dairy industry is now confronted with the problem how to ensure that the dairy products exported are in compliance with the stringent SPS requirement of that country. Inland consumer have also become conscious of food safety and quality. It is estimated that due to poor milk quality management, milk is wasted to the tune of Rs. 5500 Crores a year (Nanda and Saxena, 2000). Tackling the problem of milk quality therefore needs priority.

Present Regulatory Setup in the Country

Ten ministries and departments are administering the dairy sector. Ministry of Health deals with food safety and quality. Ten legislations in force are:

1. Prevention of Food Adulteration Act 1954 and rules.
2. Essential Commodities Act, 1955, administered by Ministry of consumers affairs by enforcing following orders:
   (*a*) Standards of Weights and Measures Act 1976
   (*b*) Consumer Protection Act, 1986
   (*c*) Infant Milk Substites Act 1992 and 1993
   (*d*) Insecticide Act 1968.
   (*e*) Export Quality Control and inspection Act 1963

(*f*) Milk and Milk Products Order, 1992

(*g*) Environment Protection Act 1986

(*h*) AGMARK Act 1937, 1986

(*i*) Bureau of Indian Standards Act, 1986

Multiplicity of laws and rules in dairy sector leads to lack of implementation and confusion in the minds of manufacturers and traders. These however, are getting merged and integrated.

The advisory and regulatory role of government towards formulation of food standards connot be questioned. Suggestions can definitely be made.

Self Imposition of Quality Regulations–Sound Alternative Milk Quality Assurance

In the counting nearly 8520 of the milk produced is handled by people who have no regard for safety and quality. Even the producer at farm level and consumer at some seldom take action to improve on this front. Consumers simply complain or grudge but considering the colossal, it is not possible for any government to solve the menace of poor quality a adulterated milk. It can be eradicated by self imposition of quality regulations at all levels or stake holders covering producer, dudhia, trader, processor, distributor consumer, sampling agencies and testing laboratories. A movement involving several NOG's universities, social organisations to educate milk, hazards for health due to its poor quality, and to clearly convince them not to buy or sell substandard not to buy or sell substandard milk. School children should be informed first to motivate their elders to shun milk of doubtful quality.

To the extent possible, all concerned should be made familiar with HACCP based quality assurance system to follow it in practice as early as possible.

Setup Private Laboratory for Testing

Important component of this porgramme, is the well equipped reliable laboratories for testing the quality. These should be managed by private organisation.

Education and Training for Food Safety

For this details should be worked out by involving competent knowledge people from government and public.

Setup Volunteer Groups

Locality wise, region – wise volunteer bodies should be formed to educate and campaign is this in a systematic planned manner. Regular testing and advising the consumer and seller to discourage transaction of unwholesome milk, should be arranged by these groups.

Lastly, arrangement for small scale milk chilling facility on reasonable charge per litre in rural milk producing area if made will go a long way in improving the quality of milk.

Time has come to act and not sleep. Similar views have also been expressed by Chawla (2002).

References

1. Chawla N.K. (2002) "Moving towards self regulation". Indian Dairyman 54(2) pp. 131-135.
2. Nanda, V. and Saxena, D.C. (2000) "Quality aspects in dairy Industry: An Indian Scenario. Indian Dairyman 52(7) pp. 15-17.
3. Rao, L.V., Rangandham, M., and Rao, B.V.R. (2002) "Quality of milk Marketed in Hyderabad City" Indian Jour. Dairy So. Sci.55(6) pp. 338-341.
4. Shourie, H.D. (1998) "Our food laws are good but enforcement is poor". The times of India dated Oct. 8, 1998, p. 11.

Chapter 16

Aspiring for a Dairy Plant: How to Survive and Prosper?

V.N. Bhargava

Prof. (Retd.) of Animal Husbandry and Dairying, B.H.U., Varanasi

Director (Operations), K.S. Singhal Dairies Pvt. Ltd., Agra

At the outset this is not a lecture, not even a talk, it is only a sharing of thoughts based on my own experience as a "hands on" executive in the dairy industry.

Normally "statistics" are a unreliable tool, because there is a famous statement "torture data long enough and it will confess to anything". Facts, however, cannot be refuted and these have to be believed, at least on the basis of what you see and learn everyday about the dairy business with your own eyes. This topic has been developed on those realities and not on statistics.

Liberalisation and Present Scenario: Quite Luring

Onset of liberalisation and delicensing programmes in 1991 attracted substantial interest in dairy sector from both Indian entrepreneurs and multinational firms based in India. Government approved more than 250 milk processing and manufacturing licenses (Abichandaui, 1994). Within few years after commencing their operations, several of these got closed or were transferred to other owners who are running these successfully.

Rapid urbanisation, growing population, rising incomes, all leading to increasing demand of convenience foods including milk and milk based products and sweets. This demonstrates enormity of existing market and vastness of the poliential that lies ahead, Waiting to be tapped.

Combine these with easy availability of finances from bauks and burgeoning three fold increase in a decade in educated unemployed, (Anon, 2005) one can easily be lured by the opportunities offered by dairy business.

"SWOT" Analysis

Any decision in any industry has to be backed by an analysis of the strengths, weakness, opportunities and threats. Let us look into these areas for the Indian Dairy Industry".

Strengths

1. *Demand profile*: beyond any doubt absolutely great and optimistic
2. *Reasonable margins*: even on liquid packed milk itself established clearly by Amrut, Indiana, Paras, Gopalji and others.
3. *Flexibility of product mix*: Tremendous. Add equipments and add on to your product lines. Milk has become an industrial fluid.
4. *Availability of raw material*: Abundant. More than 85 per cent of milk handled by unorganised sector craving for proper channelisation.

Weakness

1. *Perishability*: Pasteurization has given some relief. UHT gives it real long life. Many processes will follow.
2. *Procurement problematic*: Due to bad roads, small scattered production of milk, inadequate transport and chilling facility. With overall economic growths. These will be solved.
3. *Distribution hazardous*: Slow acting consumers and slow moving traffic makes distribution difficult.

4. *Competition*: Many small and some big people are entering this industry competition is great and has to be faced. However, market is large enough to accommodate those with fair dealings.

Opportunities

1. *Value addition and diversification*: Great scope for innovative product development, packing and presentation.
2. *Export potential*: Growing with increase in N.R.I. population in different countries. Amul is already exporting to Bangladesh, Srilanka, Nigeria, Middle East, and U.S.A.

Threats

1. *Liberalisation under W.T.O.*: Cheaper products from other dairy countries can become a big threat unless preventive measures are taken.
2. *Red tapism and corruption*: Results in slow actions and higher production costs.
3. *Milk vendors*: have strategic position in dairy industry. Efficient in disseminating wrong information and thereby doing harm both to consumers and producers.
4. *Quality of raw material*: Industry losing Rs. 1000 Crore every year due to this (Bhandair, 2003).

From the aforesaid it will not be difficult for an individual to become convinced and decide to set up a dairy plant and adopt dairy business.

To achieve success suggestions given in brief in the following pages will prove useful:

(A) For Project Planning and Implementation

1. Have a detailed project planning exercise including location and vicinity to market, availability of per cent power, water and manpower and topography of land.
2. *Milk availability*: A critical variable take into consideration potential of each village in the milk-shed, mix between cows and buffaloes yield being achieved, quality of milk, presence of traders and price fluctuations.

3. *Implementation*: For this project should be divided in phases and for each phase, period for completion should be fixed. Impletation cell should review progress at regular intervals.
4. Value engineering practices should be adopted. Expenditure should be incurred only when cost benefit factors are favourable.

(B) For Human Resource Function

1. Have proper manpower planning.
2. *Recruitment* : Phase it out have least number of well trained staff which can be gainfully employed.
3. *Optimize*: Manpower costs salaries should be linked to economic condition of the dairy plant. Employees should not remain underpaid which can result in dairy losing persons whom it needs most.
4. *Training and skill enhancement*: Should send deserving workers to better technical skills.

(C) For Milk Procurement

1. *Volume of milk*: is of great importance affecting the viability of the dairy project. Inadequate volume of procurement results not only in poor capacity utilization, but also in recovery of fixed overheads and other infrastructural costs.

   It needs to have a minimum level of procurement for generating adequate cash margins to meet day to day requirements.
2. *Price of procurement*: should vary with the selling price of milk and products with fair profit margin. Approximately 20 per cent difference should remain between flush and lean reason costs.
3. *Quality of milk procured*: constitutes the most important component of dairy business. Prevailing tendency of adulteration (adding mineral oil, vegetable oil, tallow etc.) dilution (by using water) and manipulation (supplying stored milk and buffalo milk as cow milk after dilution, adding preservative in milk and resorting to fraudulent compution of payments) calls for greatest attention for

screening the quality of milk properly. Tests and techniques are available to ensure proper quality of milk in procurement.

4. *Transport cost* : should be kept as low as possible by proper route planning, full capacity utilization of transport vehicle, proper location of milk receiving or collecting centres and so on.
5. *Regularity of payments*: irregularity in payment to milk producers/suppliers results in loss of confidence and diversion of milk to other dairy plants.
6. *Regular contact with milk producers and suppliers*: is essential for building close relationship and confidence.

(D) For Milk Processing

This demands attention to:

1. *Avoid processing inefficiency*: which results in SNF and fat lossess beyond pre-determined norms.
2. *Utility consumption*: minimize it. Strict check to be kept on consumption of utilities *viz.* power, fuel, ammonia gas, steam, packaging materials, detergents, water treatment chemicals etc. and should be in relation to output achieved.
3. *Under utilization of capacity*: may result from low milk procurement and low sale volume. This will adversely affect viability of the plant.
4. *Over creation of capacity*: becomes a big impediment in survival of dairy plant. It may be through installation of unreasonably higher capacity equipment's and civil construction much more than what is actually required. As a consequence of these, the dairy plant becomes over loaded with incidence of interest charges higher depreciation, greater costs of utilities which otherwise can be avoided.

(E) For Marketing Variables

Advertisement for sales promotion, commission to distributor and vendors, pricing and distribution or delivery of supplies are few of the salient components of marketing which are not given due consideration. Item wise relevant suggestions are made below:

1. *Marketing function*: should receive adequate attention. To exemplify : products which may be required by a group of potential customers in a particular locality are not made available in terms of quality, price, quality, time, distribution mode, packaging etc. The dairy unit expects to the customers to buy whatever is offered to them. This should not happen otherwise competitors will take advantage of this gap.
2. *Product pricing policies*: should be determined properly and kept completely, customer and plant/profit oriented.
3. *Product supplies*: should not be erratic and irregular otherwise the markets which are established after considerable efforts and expenditure then be lost.
4. *Marketing expenditure*: should not be wasteful. Do not permit unnecessary expenditure on wrong choice of distribution channel, high unproductive rates of commission on sales and sales promotion and mode of advertising. The percentage of marketing expenses to sales tend to be high, thereby impairing dairy's financial position.

(F) For Finance

Factors which need attention for efficient managing of finance of a dairy plant are given below:

1. *Accounting System*: should be such as to complete the accounts in time and does not keep arrears. It is essential to take timely remedial actions.
2. *Cost control and cash flow management*: should be efficient to exercise control on expenditure at various level and enable smooth cash flow. It is essential to have performance and cost standards established for each activity.
3. *Computerization in finance and accounting*: is essential not merely for compiling of accounts but for timely decision making. It should enable management to put tab on performance in respect of procurement trend, quality of milk received, manpower utilized, cost of utilities, sale expenses, sale receipt and so on for each month and year.

4. *Working capital*: should be properly and efficiently managed by preventing excessive purchases of stores and promptly collection of dues.

(G) Lastly Monitor Daily or Weekly Following Performance Indicators

1. Transport cost per litre of milk procured for each centre and medium wise.
2. Milk procurement growth percentage : This enables to compare its procurement with corresponding period of previous years.
3. Milk procurement as percentage of target or plant capacity.
4. Growth percentage in liquid milk sale.
5. Processing cost per litre
6. Fat loss percentage process and product wise and daily fat balance sheet of the plant.
7. SNF loss process and product wise and daily SNF-balance sheet of the plant.
8. Cost per litre of raw milk procured.
9. Overall cost/litre after including transport cost.
10. Overall cost per litre after packing and including salaries of plant staff.
11. Overall cost per litre including interest, depreciation and rent and other expenses.
12. Expenses per litre which will include administrative, office and distribution expenses per litre.

Analyse these and take corrective measures to enable your dream dairy plant to prosper smoothly.

References

1. Abichandani, H. (1994) Industrial entrepreneurs Memorandum (IEM) for milk and milk products field with Govt. of India. (Cf) Indian Dairying: Global opportunities, proceedings of XXV Dairy Industry conference, March 24-25, (pp. 59-116). New Delhi: Indian Dairy Association.

2. Anon. (2005) "17 per cent of India's graduate jobless", This Times of India dated June 23rd, 2005 New Delhi.

3. Bhandair, J.S. (2003) "Real threat to global competitiveness of Indian Dairy Industry – ADM or WTO". Indian Dairyman, Vol. 55, No. 5, pp. 53-57.

Chapter 17

Organic Milk Production: Opportunities and Challenges

V.N. Bhargava

Prof. (Retd.) of Animal Husbandry and Dairying, B.H.U., Varanasi

Director (Operations), K.S. Singhal Dairies Pvt. Ltd., Agra

Traditionally, Indians love "Nature" and every component of our civilization, style of living endeavours to bring us near to the nature. To elucidate in our music we have ragas for each season. With growth in population every thing changed. Pressure on cultivated land increased and food production system became intensively input oriented with increasing use of chemical fertilizers, pesticides, and lately use of genetically modified food items. However, now with rising disposable income, effective mass communication, growing awareness about health enviornmental and food safety, is picking up which has pushed the community to use costlier organic foods including organic milk.

Definition and Origin

Organic milk means the milk derived from organic disease-free animals, reared under natural conditions on pesticide-free feed and todder produced organically on natural fertility of soil, treated by herbal or homeopathic medicines. According to Dr. T. Nissen of,

"The Organic Service Centre Denmark", organic farming in real term is biodyanamic farming which was founded by philosopher Rudolf Striver in Germany in 1924. Organic movement, in real effective way got started in 60s based on experience of biodynamic farming. In 1981, organic farmers coordinated internationally and established the International Federation of Organic Agriculture Movements (IFOAM) (Banerjee, 2001).

The philosophy of organic farming emphasizes the need to produce food or milk in an, "integrated, humane, environmentally and economically sustainable agricultural production system" (Sem. 2002). Organic products have characteristics which differentiate them from conventional farm produces, *viz.*:

1. Integrated animal and crop production.
2. Safer products devoid of chemical residences of pesticides, autibiotics, etc.
3. More nutritious and natural products.
4. Higher quality and appearance, and
5. Environment friendly.

Organic production systems, unlike traditional systems of production, are governed by a set of standards, (organic standards) to be followed strictly by producers of organic foods. Compliance of these standards is verified by certification agencies authorized by the respective governments. Organic certification guarantees not only the quality of product but also the quality of production method or system. This more than appeal to the consumer. In the conventional products, there is no way to guarantee the production procedure.

Requirements or Standard for Organic Livestock Production

Standards for organic livestock keeping are meant to assure both an organic product to the consumer and living conditions for the animals which limit stress and promote good health. These address substances used in health care and feeding and herd management and housing:

1. Animal Management

Must fulfill the basic behavioral needs health and welfare of the livestock. Livestock farming cannot be done in isolation from crop farming. Ample provision should be made for free involvement, open fresh air, water, feed and grazing opportunities for the animals.

2. Conversion Period

The time between the start of organic management and certification of animals is known as the conversion period which is fixed as 12 months for dairy animals.

3. Brought-in-Animals

Animals should be born and raised on organic holding and not on conventional holdings brought-in-conventional animals are allowed when organic livestock is not available. When organic livestock is not available. For this situation, calves upto 4-weeks-old that have received colostrum and are fed a diet consisting mainly of full milk. Time limit for certification of these animals is 5 years and yearly 10 per cent of the animals of the same species can be replaced.

4. Breeds and Breeding

Breeds which remain well adapted to local conditions should be selected. Breeding should be natural. Artificial insemination not allowed unless it is a veterinary necessity.

5. Mutilations

Mutilations are not permitted. Following exceptions are, however, allowed viz, castration, dethroning, ringing in dairy animals and should be done with minimum suffering to the animal.

6. Animal Nutrition

Dairy stock should receive 100 per cent organically grown feeds of good quality from the farm itself. Force feeding is forbidden. In a situation, when it proves impossible to obtain feeds from organic farming the accredited certification programme may allow 15 per cent of dry matter intake to be procured from conventional sources. Synthetic growth promoters appetizers preservatives, all types of excreta feeds supplied to solvent extraction, other chemical agents, pure amino acids, genetically engineered substances are not to be added. Vitamins, trace elements and supplements shall be used from natural origin.

7. Veterinary Medicines

For -preventive health care and treatment only herbal, homeopathic medicines, acupuncture or local techniques should be used. Allopathic medicines are not permitted.

8. Transport of Animals

Journey time should not exceed 8 hours and ample water and feed should be available during the course.

9. Record Keeping

Producer should keep written records concerning production and processing. Input used sources of inputs, medicines given, sources of feeds and todder given and welfare measures taken are to be recorded. Besides, the records should contain (i) Third party certification (ii) Audit trails (iii) annual inspection (iv) material list (iv) conversion periods (vii) sustainable farm plan.

10. Labour

No child labour to be used in production, processing and marketing of dairy products.

Scope for Organic Milk Production

One can witness people daily going in the morning to nearly Khatals in their localities to fetch milk dram in their presence. They want to ensure that the milk their family consumes is unadulterated and are ready to pay premium price. Right now, they fail to notice what is fed to the animals, under what is conditions the milking of the animal is done and the milk is containing antibiolies etc. However, awareness is growing; and sooner demand for organic milk will appear as is already there for vegetables, fruits and food products. A 'swot analyses" can throw better light about the unfolding market for Organic Milk.

SWOT Analysis

(A) Strength

1. *Indigenous cattle and buffalo wealth*: India is enriched with an enormous bovine population (273.3 million) with 26 native cattle breeds (10 per cent of worlds) all breeds (8 breeds) of riverine buffaloes producing over 84 million

tonnes of milk annually making number one milk producer in the globe.

Milk production is predominately the domain of small holding farmer in a mixed farming system of which 70 per cent constitute dry land or rain fed areas. As such use of chemical fertilizer and agrochemical is very low.

Indigenous breeds have unique characteristics like endurance, docility, resistance to tropical diseases, ability, resistance to tropical diseases, ability to utilize coarse todder and heat tolerance. This is why these have attracted attention of foreign countries particularly south America. Programmes have been taken up to develop these (Chander and Kumar 2002). Energy efficiency of desi Cattle is 17 per cent as compared to 4 per cent of American beef breeds (Chander and Kumar Loc. cited). As such these are most suited for organic farming.

2. Enough dung an important component of organic farming is manuring. Yearly Cattle dung, in India has a fuel value equivalent to 35 million tonnes of coal. An estimated 1/3 of this or 300 million tonnes of drng is used as fuel which can be diverted for manuring if bottled methane gas from bio-gas plants is made available to such families (Anon, 2005)
3. *Expanding market*: With growing awareness for environment, chemical free foods, health and nutrition, market for organic foods is opening up.
4. *Rising income*: Per capita income is continuously showing an upward trend. During 2004-2005, percapita has gone up by 10.7 per cent (Anona 2005). Consequently, the quality consciousness and willingness to pay more for good quality products including organic items would definitely go up in coming years.
5. *Export potential*: The market for organic dairy products in Austria, U.K. and Denmark is annually growing @ 30 per cent whereas in sweden, Germany and France @ 15 per cent. Besides, organic dairy products are fetching 60 per cent greater price than conventional dairy products (Banerjee, 2001). Organic dairy farming add up greatly to the export item list of Indian dairy products.

6. *Contract farming system*: Loud holding are small. With the introduction of contract farming system, this bottle neck can be removed. Larger organic dairy farms will them not remain a distant dream.
7. Treasure chest of herbal, Ayurvedic and Homeopathic medicines. An important prerequisites for producing organic milk.
8. *Milks is cost competitive*: Indian milk, cost-wise is quite competitive being stands fourth lowest in the world and fortunately these are no subsides in the Indian dairy industry (Jain, 2002). This is a great plus point.

(B) Weakness

1. Seventy percent of the milk producers are laudless. Most of them re-illiterate. These will be major landscapes for organic farming. Contract farming and present large number of educated unemployed, put together promise a bright ray of hope to face it.
2. Lack of training in organic farming. In due course facilities will be created to take care of this.
3. *Herbal and homeopathic remedies* : Available at present are of unreliable efficacy.
4. Long gestation period required for start ups.
5. *Use a child labour*: No child labour is permitted in organic dairy farming. This may be yet aware constraint for India where children share a lot of work relative to livestock production. As incomes grow in rural areas, this handicap will be nullified.
6. *Clean drinking water and green fodder*: Are to be supplied to animals in organic milk production. Both these requirements appear difficult to be met. Adoption of contract organic dairy farming will certainly erase these limitations. Moreover, our breeds have strong survival poential and can maintain themselves on crop residues.
7. *Shrinking Land and grazing area*: Due to population growth land and grazing areas are shrinking due diversion for crop farming. Looking to the potential and to efforts underway to maximize production of food crops, in near

future requirement for grazing area certainly can be fulfilled.

(C) Opportunities

1. *High profit margin*: On an average organic milk produces are charging 60 per cent higher price than the milk produced by the conventional method. (Banerjee, 2001)
2. *Export potential*: Export of livestock products is fast growing. Export of dairy and poultry products during 2004-2005 period recorded 65 per cent increase over the previous year in terms of dollars (Anon., 2005).
3. *Value addition and diversification*: As in the conventional milk production and processing, organic milk processing and manufacturing also offers a great scope in this respect. Add to the equipment line and increase your organic milk – product lines. With value addition, profits will move faster.

(D) Threats

1. *Multi-agencies involvement*: For certifications and labelling, organic diary production has to face several agencies. Process becomes slow and the path becomes difficult to tread and tackle.
2. *Lack of infrastructure and bureaucratic slow action*: Formation of committees and agencies for certification and accreditation's are to be formed for different regions and stages by the government. This remains slow and tiring for the organic producer. However, as the things get moving, the process will catch up. Successful implementation and efficient guidelines can remove these constraints.

   Looking to the aforesaid 'SWOT' analysis, it is not difficult to conclude that 'strength' and opportunities for out weight 'weakness' and threats. The scope for organic milk industry is great.

Opportunities and Challenges

Haring established the scope of organic dainging in the country let us briefly look to the opportunities it offers:

1. Opportunities in production of Organic feed and fooder.
2. Opportunities in raising organic replacement livestocks.
3. Opportunities cultivation processing of herbal and ayurvedic medicinal plants.
4. Opportunities in preparation of homeopathic, ayurvedic and herbal medicines for livestock. Besides local market big export market is true for these.
5. Opportunities in processing and production of organic dairy products.
6. Opportunities in promotion and presentation of organic dairy products.

The only main challenge is how fast and successfully programme for organic milk production is implemented and carried out efficiently.

As such, let us not stand still, but join hands, to work shoulder to shoulder to put India firmly on top of the map of "organic" products.

When Brazil and Argentina can achieve why not we?

We have the resources, what we need is the "Will" to do and we will succeed.

References

1. Anon (2005) "Bottled biogas in Gorakhur village" In Time of India, Delhi (ed) dated 6th June, 2005, p 4.
2. Anon (a) (2005) "You grew 11 per cent richer in 2004-05". In the Times of India Delhi (ed) dated the 2nd July, 2005, p7.
3. Anon (b) (2005) "Krishi Sambandit Utpacton Ke Nriyat Me Record Badhotri". In "Dainik Jagran", Agra (ed) dated 4th July, p. 13.
4. Banerjee, A (2001) "The organic dairy and Food Industry – A paradigm shift", Indian Dairyman 53 (12) pp. 31-33.
5. Chander, M. and Kumar, S. (2002) "Indigenous Cattle and buffalo Wealth of India: Exploring its role in promoting brganic farming practices". In Proc. of the National Workshop (26-27 Nov, 2002) on, "Organic Animal Husbandry Standards", I.V.R.I., Izatnagar-243122 (U.P.) India.

6. Jain, D. (2002) 'Global opportunities for Indian dairy industry" In Indian Dairyman 54, (2) pp. 70-71.
7. Sen, A. (2002) "Standards of Livestock Farming: A Comparative review" in National Workshop on Organic Animal Husbandry (Loc. cited).

Chapter 18

Enrichment of Crop Residues and Low Grade Roughages for Milk Production in Dairy Animals

A.B. Mojumdar

Plant Animal Relationship Division, Indian Grassland and Fodder Research Institute, Jhansi – 284 003, U.P.

Livestock forms the integral part of agriculture since ancient times and its contribution to the national economy through livestock production like milk, meat and wool is invaluable. Adequate availability of such animal products is hampered by the non-supply of quality feed. Supply of adequate nutritious feed resources for enhanced and sustained livestock production is most important aspect for the successful exploitation of this sector. It is paradoxical than shortage of animal feed and fodder is yet continuing.

Both quantitatively and qualitatively there exists a large gap in demand and supply offered resources. There are several projections which shuns continuous deficit of fodder and livestock feed in coming years. There is deficiency of about 25 per cent straw, 30 per cent green fodder and 47 per cent concentrate in our country. In India, wheat, rice and coarse grains like maize, millet and sorghum form the bulk of agriculture crops grains in India and therefore low-grade roughages are the principals some of livestock ration. The gap between the supply and demand for quality animal feed continuous to enlarge owing to various constrains like availability of land and resources input. The grazing land as well as lands put under fodder production are shrinking as a result of competing demands for food under ever increasing population pressure, organization and industrialization. However, there is shortage of green fodder round the year particularly during scarcity and lean period. In India, area under cultivated fodder, which is 8.33 m ha that is not likely to increase tangibly rather may decrease due to competition with other agriculture crops. By and large, the ruminants particularly cattle and buffaloes are forced to thrive in low quality roughages like crop residues, straws, dry stalks, stubble and dry grasses available for wastelands.

Quality of Low Grade Roughages

Straw, crop residues and dry grasses from the bulk ration for livestock feeding in many parts of word especially in tropical and sub-tropical regions. These are very poor in quality and are characterized by low crude protein, minerals and high fiber and lignin contents (Mojumdar, 2002). Thus, its voluntary intake and digestibility is also adversely affected when fed to animals. Sole feeding of such roughages even fail to cater the maintenance requirement of ruminants causing adverse animal performance. Increased straw quality is a priority when objectives of animal production is high and supplements like concentrate mixture are expensive and hardly available. Therefore, there has been an increased interest in feed processing and enrichment of these low-grade roughages for improving it nutritive value. Low-grade roughages can be divided into different categories based upon their physical and chemical characteristics.

CP content of these low grade roughages are very low and much below than the critical level of 7 per cent which is necessary to meet the maintenance requirement of the ruminants.as well as necessary for sustaining conducive environment in rumen. The rate of passage of these roughages is also low due to higher functional gravity of particles.

Ways of Enrichment of Low Grade Roughages

Several ways for enrichment of such poor grade roughages have been worked out which can be classified into chiefly three groups:

1. Physical methods
2. Chemical methods
3. Biological methods

Physical Methods

Physical methods like chaffing, grinding, pelleting, steam treatment, water soaking and radiation or other alterations of the feed ingredient are practiced to enhance the nutritive value of such feeds.

Chopping

Crop residues/straw may be processed to alter the physical form of particle size. Chaffing is of utmost importance and it directly linked to intake. Chopping has been found to cut down the particle size, which ultimately increases the intake and digestibility of straw. Consumption of chopped rice straw has been found higher than of long straw (Costillo *et al.*, 1982). Chopping increases the rate of passage and better fermentation in the rumen (Owen, 1984). Chopping also reduces the wastage (Doyle, 1986).

Grinding and Pelleting

Grinding and pelleting increases voluntary intake. It has been found that finer the grind, the more rapid is the rate of passage and higher is the intake. Grinding and pelleting improve uniformity and density of feed and thus increases the surface area and thereby expedite free accessibility for ruminal microbial enzymes.

Several advantages such as:

1. Improved bulk density
2. Better handling Removal of antiquality organism
3. Reduced selective feeding
4. Improved animal performance
5. Decreased feed wastage

However, the cost of pelleting is of greater concern than for most other processing methods.

Water Soaking

Water soaking affects the internal physico-chemical characteristics of straw through swelling of fiber, softening of particles, loosening the linkages with structural carbohydrates which ultimately increases the palatability. Cattle and buffaloes receiving water soaked wheat straw has been reported to consume more straw dry matter (2.9 per cent) than unsoaked (2.6 per cent) if straw is soaked 10-12 hrs before feeding. Improved intake, digestibility has been reported if wheat straw is soaked with water overnight before feeding

Table 18.1: Effect of Soaking in Wheat Straw

	Intake%	*DMD%*	*ADF Digestibility%*
Dry Straw	1.88	52.2	40.8
Soaked straw	2.18	55.8	45.3

Sources: Mathura, M.C. and Sharma, V.V., 1985. Indian J. Animal Sci. 55: 218 – 220

Steam Treatment

Steam treatment of low quality roughages has been found to increase voluntary intake and digestibility in cattle by degradation of cellulose and hemicellulose. Steam treatment of sugarcane bagasse, paddy straw and sorghum stover for 30 to 60 minutes has been reported to reduce hemicellulose and increase VFA production.

Chemical Methods of Enrichment

Alkali Treatment

In the past, most widely applied chemical used for improvement of low-grade roughage, has been sodium hydroxide. It is a powerful swelling agent of the amorphous region of cellulose. It breaks down lignin-cellulose, hemi-cellulose linkages in the cell wall and increases the cellulose digestibility. Sodium hydroxide is most effective delignifying agent.

50g Sodium hydroxide is dissolved in 0.5 to 2.0 litre solution per kg of dry matter of roughage, this is sprinkled over the straw. At this rate of treatment, batches of straw are prepared before feeding. However, the use of sodium hydroxide has been found corrosive. The use of ammonium hydroxide attracted the attention as it also increases the N content in addition to intake and digestibility. Because of the requirement of special device for application of the liquid ammonia and low environmental temperature, this technique is not feasible in countries like India and other tropical and sub tropical regions of world. However, this technique is being followed in temperate countries like Europe. Calcium hydroxide has also been tried out but due to its low solubility, imbalance of Ca: P ratio and being a weak alkali, it has not been found to be very effective.

Urea Treatment

In tropical countries like India, ammoniation through urea treatment has been found to be the most important way for enrichment of low-grade roughages because of its ready availability and application. In this process, lot of work has been done. Roughage is treated with 4 per cent urea and 40 per cent water and is sealed for a month to allow the generation of ammonia from urea. Higher OM intake and OMD observed when 4 per cent urea 40 per cent water treated dry grass were fed to buffalo heifers. It has been further observed that feeding of 3 per cent urea and 30 per cent water treated wheat straw to lactating cows increased OM intake by 13.92 per cent, CP content by 2.4 fold improvement (2.94 per cent to 7.13 per cent CP), higher OM digestibility, DCP and TDN intake. Milk urea was though higher in the cows fed with ammoniated straw but was within physiological range. No adverse effect is found on the blood profile such as serum protein, albumin, globulin, glucose and urea level by feeding ammoniated straw. Enrichment of wheat straw through 3 per cent urea addition was able to economise the concentrate mixture requirement to the time of 35 per cent in the diet of lactating cows without affecting milk yield, milk composition and health performance of the animal (Mojumdar *et al.*, 2002). Urea treated straw fed animals has been found to possess high CF digestibility which may be attributed to the increase solubilization of structural constituents as well as intensive microbial fermentation that takes place from a substantial release of added nitrogen to the rumen.

The response to urea treatment varies from crop to crop and between the slender and coarse stemmed straws. In thin stemmed straws, maximum response to treatment has been reported in the order of wheat, barley, rice and in thick stemmed straws sorghum (finger millet) maize stover. The straws having more poorer in quality and feeding value had more response of ammoniation. Similarly, more response of ammoniation has been observed in dry grass than millet and oat straw. This technology of urea treatment has now been well perfected and needs its adaptability by the farmers as crop residues from wheat and paddy are available in plenty. Yadav (1999) after feeding 2 per cent urea + 4 per cent lime treated paddy straw to the growing buffaloe calves observed significant increase in paddy weight gain (477 g/h/day vs 373 g/h/d).

Oxidising Agents

Oxidising agents cause lignin solubilization and thus improve quality of low grade roughages. Bleaching powder has been found equally effective as sodium hydroxide cleaving out lignin but at a higher level (beyond 2 per cent level of treatment), the residual chloride in treated straw has been found toxic to rumen bacteria and thus digestibility has been inhibited. Sodium sulfide has been found more effective than sodium sulphite. A number of other oxidising agents like potassium permanganate and alkaline nitro-benzene have been used for enrichment of straws but they have proved to be too expensive or toxic.

Biological Treatment

The use of microorganism to enrich roughage quality has also been found an alternative to physical or chemical treatments. lignolytic fungi selectively degrade lignin and leave behind better digestibility cell wall carbohydrates for further fermentation and digestion in the rumen. The results of fungal treatment of ammoniated straws have shown that the process is accompanied by loss of organic matter, reduced digestibility, increased DCP and reduced TDN in the fermented product causing a net loss of nutrients, the conversion of roughage into a better quality feed by delignification through white rot fungi aims at maximum lignin degradation with minimum degradation of cellulose and hemicellulose. The two stages "Karnal Process" involves use of *Coprinus fimetarius* fungus which captures nitrogen compounds generated by urea in the first stage of treatment.

Supplementation of Urea-Molasses-Mineral Block Licks (UMMB)

Supplementation of urea-molasses-mineral block with low grade roughages improve the nutrient utilization and economise the concentrate requirement by about 25 per cent for animal yielding 4-5 kg milk per day.

References

1. Casillo, L.S., Roxas, D.B., Chavez, M.A, Momongan, V.G. and Ranjhan, S.K. (1982). the effect of concentrate supplement, chopping and soaking rice straw on its voluntary intake by carabaos. pp 74-80. In "the utilization of fibrous agricultural residues as animal feed" ed. P.T. Doyle, university of Melbourne.

2. Jodha, N.S. (1986). Common property resources and rural poor in dry regions of India. Economics and Political Weekly, 21 (27), 1169 - 81.

3. Mojumdar, AB., Das, M.M., Maity, S.B. and Anil Kumar (2002). Nutrient utilization and performance of crossbred cows fed urea treated straw based diet. Animal Nutrition and Feed Technology.

4. Yadav, K.K. (1999). Effect of urea and lime treatment in paddy straw on chemical composition and nutrient utilization in growing calves. Ph.D. Thesis. H.AU. Hissar.

Chapter 19

Fodder Bank Concept for Feeding Dairy Animals during Scarcity

A.B. Mojumdar

Plant Animal Relationship Division,
Indian Grassland and Fodder Research Institute,
Jhansi – 284 003, U.P.

Drought of 2002-03 affected 62 per cent of bovine population in 18 states of the country. Due to inadequate nutritional supply during scarcity, milk production reduced by 7 per cent in Tamilnadu, 8 per cent in Maharashtra and 22 per cent in Rajasthan. Poor nutritional support also caused adverse effects on dairy animals which included poor fertility and reproductive function, bovine population, breeding cover and draught energy. The available fodder resources may be effectively utilized through ease transportation of fodder resources from surplus places to deficit places. There is therefore an urgent need to establish fodder bank in surplus fodder zones and evolve methods for easy and economic transportation to deficit areas during scarcity. The establishment of fodder bank will also provide nutritions fodder to dairy animals round the year.

The available feed resources need to be effectively utilized as there already exist a deficiency in both dry as well as green fodder alongwith concentrate. Many farmers in surplus areas are still in practice of burning the crop residues which needs to stored as fodder bank and be supplied to scarcity areas after proper densification. Storage of fodder as well as its transportation is a problem for fodder growers, owing to its large bulk density. Efforts are required to find the ways for reducing the bulk density and densification and block making are the real answer for developing fodder bank concept.

Agriculture crop residues are the major roughage source to the dairy animals. These are bulky and difficult to transport from surplus to scarcity regions. Therefore, densification plays vital role to reduce the transportation cost by lowering the bulk density. Moreover, the crop residues are deficient in several nutrients particularly CP and mineral and are not even capable to fulfill the maintenance requirement of adult ruminants. The densification of roughages provides ample scope for incorporation of unconventional or top feeds. Thus, all out efforts are essential to popularize this technology through developmental agencies such as cooperatives, NGO and feed manufacturers etc. so that livestock owners can accrue benefits by feeding the balanced complete feed blocks to their animals.

Storage of Complete Feed Blocks

Complete feed blocks were manufactured and stored in conventional godown under natural climatic conditions for one year. The temperature ranged from 10 to 42°C and relative humidity ranged from 20 to 100 per cent . The samples were taken every month and were analyzed for CP, DM, total ash and organic matter contents. The blocks were regularly inspected for growth of fungus.

The changes became prominent after 3 months of storage, the crude protein content started decreasing slowly and gradually and decreased by 1.10 per cent at the end of the year, the total ash content of feed started increasing and increased by 2.48 per cent, whereas the organic matter contents started decreasing and decreased by 2.49 per cent at the end of the year.

Overall physical appearance of the feed blocks showed no change upto one year of storage. The smell was also normal. There appeared no visible growth of fungus during the storage period.

In vivo Evaluation of Feed Blocks

In an experiment, the adult buffaloes were fed complete feed blocks made of four similar rations except the type and nature of straw. The DM intake was more in paddy straw based blocks as compared to wheat straw based block. Grinding of straws improved its DM intake. The digestibility co-efficients of all the nutrients were higher in paddy straw than wheat straw based blocks and the co-efficients decreased due to grinding of straws.

Table 19.1: DM Intake and Digestibility Co-efficients in Buffaloes

Parameters	*Complete Feed Blocks*			
	Wheat Straw Unground	*Wheat Straw Ground*	*Paddy Straw Unground*	*Paddy Straw Ground*
DM intake/ 100 kg body weight	1.90	2.26	2.36	2.46
	Digestibility co-efficients (per cent)			
DM	50.69	49.49	58.25	55.02
CP	66.59	64.52	69.10	64.99
CF	56.00	54.02	65.54	63.89
EE	53.74	49.00	59.52	54.56
NFE	54.14	52.10	60.00	58.96

In another experiment, growing buffalo calves fed on green forage based complete feed blocks (wheat straw, berseem, molasses, urea, mineral mixture and salt) and conventional feed (wheat straw, green fodder and concentrate mixture). The dry matter, nutrient digestibility, nutritive value and body weight gain was significantly higher in second than the first group, but, the cost per kg weight gain was Rs.3.58 more in group fed on conventional feed. The results indicated that green fodder based complete feed blocks can be developed after wilting (20-25 per cent moisture) and can be fed to the animals safely.

Jaglan (2002) conducted an experiment on crossbred heifers by feeding complete feed blocks containing urea treated wheat straw (T_1). Wheat straw was replaced by urea treated paddy straw (T_2), mustard straw (T_3) and sugarcane bagasse (T_4) at 50 percent level, respectively (Table). In each treatment roughage to concentrate ratio was kept 60:40. The DM intake and daily body weight gain were higher, while the cost per kg live weight gain was minimum in T_1 than other treatments. The results showed that urea treated crop residues can successfully replace urea treated wheat straw at 50 per cent level in complete feed blocks with a daily gain over 530 g/day.

In another experiment on lactating crossbred cattle, Sunda, (2004) studied the effect of 50 per cent replacement of wheat straw (T_1) with paddy straw (T_2), sugarcane bagasse (T_3) and mustard straw (T_4) on milk yield and its composition. The roughage to concentrate ratio was maintained 60:40. The digestibilities for all the nutrients were higher in T_1 and T_2 as compared to T_3 and T_4 treatments. The milk yield significantly ($P>0.05$) reduced when 50 per cent of the wheat straw was replaced by sugarcane bagasse or mustard straw, however, the treatments did not affect the milk composition.

Table 19.2: Effect of Different Sources of Crop Residues on Milk Yield and its Composition in Cattle

Parameters	*Treatment*			
	T_1	T_2	T_3	T_4
	Digestibility co-efficients (per cent)			
DM	63.36^b	63.98^b	57.45^a	58.08^a
CF	60.21^b	68.84^b	56.36^a	56.50^a
NFE	65.96	64.65	63.00	62.80
Milk yield (kg/animal/day)	10.80	10.87	10.15	10.02
	Milk composition (per cent)			
Fat	4.18	4.0	4.49	4.36
Total solids	12.36	11.84	12.56	11.68
SNF	8.54	8.24	8.52	7.86
CP	3.57	3.54	3.48	3.46

Benefits of Densified Feed Block

1. Reduces the cost of transportation of cereal straws and will also minimize the risks of road accidents, because of pregnant trucks.
2. Minimizes the cost of storage of cereal straws and thus will help in solving the problem of acute shortage of storage space, particularly in urban and peri urban areas.
3. The feeding of balanced complete feed blocks saves time, labour and reduces the problem of nutrients deficiency and wastage of feeds.

4. The feed blocks are cheap an more palatable and digestible as compared to pelleted feeds as well as conventional feeding of straws and concentrates, fed separately. Thus the technology developed increases animal performance and will curtail the cost of livestock production.
5. The technology will be particularly helpful in solving the problem of feeding livestock in floods/droughts affected areas.
6. In case of complete feed blocks manufactured by this technology, there is no need to grind the roughage portion as required in feed pelleting and thus the digestibility of the feed is not affected.
7. The machinery and equipment required is not so costly as in case of pelleting.

Complete feed blocks can be manufactured using all type of conventional crop residues with particle size 2.5 to 3.0 cm. For preparing good quality blocks, molasses to be added @ 15 per cent and moisture @ 15 to 17 per cent of the material on dry matter basis. Concentrate mixture can be included up to any level and green fodder as berseem upto 70 per cent of the total material after wilting for 96 hrs (17-20 per cent moisture). The material can be densified to 2.5 to 4 times, when pressed at 350 to 400 kg/cm^2 for 5 minutes as dwell time. Densification reduces the cost of transportation and storage. The product is exclusively important in flood/drought conditions. During scarcity, unconventional feed ingredients can be easily incorporated into the feed blocks. The feeding of densified feed results in to improvement in animal performance.

Limitations/Constraints in Commercialization

1. 'Ready to use' plant is not available in the market, it has to be got fabricated and standardized.
2. Commercial production of complete feed blocks at large scale will require systematic large-scale commercial plant.
3. Huge infrastructure facilities are required to handle large quantities of feed ingredients, particularly low-density roughages and complete feed blocks.

To combat the feed scarcity situation during famine such as drought, which is regular feature in arid regions of India. The surplus feed and fodders from different parts of country can be used following block technology and be transported to required places. For example– plenty of rice straw is being produced in northern and central part of India, states likes UP, Haryana and Punjab. Rice straw is not being used for animal feeding in these states because good quality fodder availability in states are adequate. After grain harvesting rice straw is being burned in the field to grow rabi crops. This rice straw can be collected free of charge and blending with some concentrate ingredients complete feed blocks can be prepared, easily transported and stored for future use. Since keeping quality to these blocks is quite good hence fodder banks can be developed.

Complete feed block technology seems promising for intensive animal production systems through the improvement of rumen function by optimal rumen fermentation via the adequate and continuous supply of desired nutrients thereby minimizing fermentation and absorptive losses. Block based feeding strategies are pertinent under urban animal production systems due to less space requirement for storage and ease in transportation and feeding. Low transportation cost, high keeping quality and less residual losses will make the technology economically viable for animal producers. Block technology will helps in developing low cost feed formulations, avoiding refusal of unpalatable portion of plant residues, improves utilization of NPN compounds that enables use of locally available ingredients in an efficient manner.

Reference

Jaglan, B.S. 2002. Urea teated crop residues based complete feed blocks for growing animal. M.Sc. Thesis submitted to HAU, Hissar.

Chapter 20

Conservation of Surplus Fodder and Grasses for Feeding Dairy Animals during Lean Period

A.B. Mojumdar

Plant Animal Relationship Division,
Indian Grassland and Fodder Research Institute,
Jhansi – 284 003, U.P.

Surplus green herbage at the flush growth period is conserved either as hay or silage or artificial dehydration for its feeding to livestock during lean period when availability of fresh forage is meagre or negligible (mid October-mid December and mid April-June). Indian being a tropical monsoon bound country, large quantity of surplus kharif forages are produced which are frequently excess of needs. It is therefore, essential to preserve the excess forage both during kharif and rabi season at proper stage of maturity to provide nutrients during lean period. Ensilage has many advantages over the other methods of forage conservation. There is high-energy output in high temperature dehydration. Hay making is difficult during monsoon season because of unschedule rainfall and little availability of sunlight in addition to causing leaching of nutrients. Crops can be well conserved as silage but it requires precautions as lack of understanding of the factors associated with ensiling technique may produce silage of bad quality leading to poor animal performance.

Principle

Silage is the material produced by controlled fermentation of crops under anaerobic condition. This process is known as ensiiage and the container used for the purpose is known as silo.

When herbage is put in an airtight container, naturally occuring bacteria ferment the carbohydrates (sugar) present in the herbage to produce mainly the lactic acid. The aim is to achieve a sufficient concentration of lactic acid to prevent other type of bacterial activity *i.e.* clostridial activity. The main requirements are the exclusion of air to maintain anaerobic conditions and to discourage clostridial fermentation which leads to production of carbon-di-oxide, ammonia, amines, butyric acid etc.

Ways of Controlling Fermentation

There are several ways in which the clostridial type fermentation can be checked. One of the commonly practices way is to increase the dry matter of the herbage. In ensiled crops containing 30 per cent or more dry matter, crop ensile satisfactorily. Promotion of lactic acid fermentation is important pathway for controlling clostridial growth. Other ways of controlling fermentation, is by addition of preservatives and additives.

Microbial Activity

As anaerobic conditions are achieved in silo, the species of *Escherichia, Bacillus, Clostridium, Leuconostoc, Lactobacillus* and *Pediococcus* develops. Lactic acid bacteria *(Streptococcus, Leuconostoc, Lactobacillus* and *Pediococcus)* are the important for preservation of silage of good quality.

Silos

The following type of silos are generally used:

1. Pit silo
2. Tower silo
3. French cum bunker
4. Trenches
5. Drum and PVC silo.

The silo must provide a solid surface to permit consolidation of the ensiled material and elimination of air. It must protect the silage from water. In India, pit silo is most common.

Techniques of Silage Making

1. Dry Matter

Dry matter should be above 30 per cent. Crops of high moisture should be ensiled with the addition of preservative and additives. In poor weather, avoid wilting and use additive for good fermentation.

2. Stage of Growth

Crops should be cut at a proper stage of maturity as it is the most important factor for controlling the silage quality.

Sorghum Maize Oat: Flowering to dough stage Milk to dough stage.

Grasses: 50 per cent flowering to dough stage early flowering stage.

3. Chop, if Possible

Crop should be chopped before ensiling. For good silage, the shorter the chop length, the better the quality. Chopped silage is more palatable to livestock and little chance of secondary fermentation.

4. Filling

Silo should be filled rapidly and avoid leaving it open. Seal as soon as possible. Packing is important to get anaerobic conditions. It should be hardly pressed to that no air pocket is left in the silo otherwise chances of mould formation will be there which will spoil the silage. After filling, silo should be covered with polythene sheet and followed by layer of soil etc.

5. Removal

After 45 days of ensilment, silage can be removed for it's feeding to animals. Care should be taken in removing the silage from silo. It should not be allowed to deteriorate after the silo is opened for feeding. Covers should be kept firmly in place as long as possible and the minimum face should be exposed at one time. The sugars, proteins and lactic acid present in the silage are subject to attack by mould growth and oxidation as some as air is allowed to fermentation and causes loss of feeding value and intake by the animals.

6. Silage Quality

Silage quality is determined by mainly the odour, physical state, pH, ammonia nitrogen, volatile acids and lactic acid.

It should be of pleasant smell and dry in nature.

pH : 4.5 to 5.

Ammonical nitrogen of total N : Less than 10 per cent of total N

Butyric acid - less than 0.2 per cent

Lactic acid - 3 to 12 per cent

Volatile acids, alcohol should be low.

Hay Making

Conservation of high quality forages by drying is termed as hay making. In India, sunlight is available in abundance, which enables farmers to dry the green forage in open sunlight and thus making hay more economically. The hay making leads to reduction of moisture content to 10-20 per cent which inhibits the enzyme activity in the plant to be conserved. Thin stemmed cereal crops like oat, range grasses, guinea grass, range legumes particularly Stylosanthus, Siratro, Dolichos, all the cultivated legume fodder like berseem, lucerne cowpea and jowar are suitable for hay making. Legumes because of their high buffering action and high nitrogen content, are and therefore these legume forages should be more conveniently conserved as hay.

Hay making is relatively more convenient and easy for Indian farmers. It can be done by sun-drying. During inclement weather, hay making may be done by drying under shed or artificial drying through solar energy. Harvested forage particular! y thick stemmed should be chopped and spread over the ground for sun curing and layers should be changed every day to prevent any sort of fermentation or bacterial growth. After, it is well dried (dry matter content at the time storing should be around 85-90 per cent), this can be stored for feeding during lean petiod. Thin stemmed crops and legumes can be dried without chopping while thick and stemmed fodder like sorghum, maize and bajra require chopping of crushing before they are allowed to dry.

Factors Affecting the Quality of Hay

1. Plant species
2. Stage of harvest
3. Leaf stem ratio
4. Chemical composition
5. Physical form
6. Deterioration during storage

Methods of Haymaking

1. Hay Curing Sheds

This method is more suitable during rainy season. It can be more economically constructed by using locally available material like jungle wood and munj coir and a netted structure. The layout of the shed should be such that it's length could run across the usual direction of wind. The hay prepared by this method posses bright green colour, higher CP content and pleasant smell than the sun cured hay.

2. Gable Shaped Structure

This type of hay making device hay been developed at IGFRI, Jhansi with the help of galvanized woven wire fencing material (width 120 cm, dia of wire 2.4 mm. and mesh size 33 cm x 14 cm). In this method, a number of rows of angle iron of 115 cm height with consecutive post 350 cm apart and rows of posts 350 cm apart are erected on the ground. Each post has a hole and bored near the free end. Two rolls of wooven wire fencing material are spread along each row of posts in such a way that their common edges which formed an apex are tied at intervals near the free end of the posts. The free edge of each roll is fixed to the ground at intervals by means of small wooden pegs. For drying every quintal of freshly harvested berseem, about 6 sq. meter of fencing material is required. This device provides a well ventilated slopping support for drying of fodder. The fodder is spread uniformly on the frame, on the both side of slope. The plant materials are turned once in a day. Normally it takes about four days of time for making 2nd cut berseem hay, while in case of 3rd cut berseem, drying is completed in 24 hrs only. In order to minimize shedding of dry leaves, the hay is transferred to hay barn. The loss of nutrients is greatly reduced by this method and ensures rapid drying. This method is particularly suitable for berseem and lucerne hay.

Chapter 21

Meeting Dairy Farm Requirement: Production and Conservation of Green Fodder

S.K. Kaushish

Central Arid Zone Research Institute,

Jodhpur – 342 003

India, with merely 2.4 per cent of world's geographical area, supports over 15 per cent human population, nearly 16 per cent cattle population, 55 per cent buffaloes, 4 per cent sheep, 20 per cent goats and 1 per cent pigs of the world. In addition, 1 m camels and 1.9 m ponies are also supported. Crop and livestock are two functional components of farming system, which determine agricultural economy and dairy development. An integrated approach considering both the animal and its ration is essential to increase and stabilize dairy production in our country. With the adoption of large scale cross-breeding programme for cattle using foreign breeds like Jersey Holstein, Red Dane and Friesian, the average milk yield from the cross breeds is likely to go up to 3000-3500 liters per lactation as against 1000-1200 liters from the local breeds. There cross breed animals would require about 30-40 per cent more green forage than the indigenous and therefore, the supply of sufficient quantity of nutritious green fodder is of vital importance. There is a large gap between demand and supply of green fodder, the production being around 574 m t against the total requirement of 745 m t it is because of the fact that only 4-5 per cent (about 8 m ha) of the total cultivated area in the country is allocated to forage production where as for success of a mixed farming system at least 10 per cent of each holiday should be used to produce green fodder to make significant contribution to the farm income from complementary dairy enterprise, specially during the lean periods of forage and cash crop rotation. In some state of country like Haryana Punjab Western U.P. part of Gujarat and Maharastra, fodder production is either sufficient or surplus while in other parts including Rajasthan the fodder availability is deficit. However, increasing pressure of human population and consequently rising demand for food grain it is difficult to divert more land to forage production especially in the densely populated state. The solution, therefore, lies in maximizing forage production in space and time through intensive cropping system with the improved production technology.

The low productivity of Indian livestock is mainly due to chronic shortage of feeds and fodder coupled with poor quality. The emphasis on roughage as basis for feeding leads to other points of importance, the often-low quality forage, high variability of quantity, conservation problems etc. Flush season (October to March) abundance and lean season (April to September) scarcity are the characteristic feature of Indian dairy, which is closely related to the green forage availability.

Fodder Production Systems

Intensive Forage Production System

Considering both the dairy farmers who requires technology of forage production round the year and the small farmers who own a few herds of milch animal and require cropping systems to obtain maximum forage yield per unit area and where irrigation facilities are available a farmer can earn better by raising fodder crops through out the year. Forage production from cultivated area much depends upon the type of crops grown and the improved varieties. Generally, perennial forage N-B hybrids gave higher assured production in most of the locations with highest green fodder yield. Crop sequences have been developed to achieve high yields of nutritious green fodder along with the maintenance of and productivity over a long period of time. Pearl millet-Lucerne-sorghum, Pearl millet-oats-sorghum and Pearl millet-Lucerne-maize sequences are found superior. Some important crop sequences and their forage production potential for different zone in the country are given (Table 21.1)

Table 21.1: Fodder Production Potentials of Improved Crop Rotations

Sl.No.	Crop Rotations	Green Fodder Yield (t ha^{-1})
	I. Northern Region	
1.	White clover + rye grass	131
2.	N-B hybrid+velvet bean-berseem+sarson	123
3.	Deenanath grass-berseem-maize+cowpea	200
4.	N-B hybrid + sundal	140
	II. Central and Western Regions	
1.	N-B hybrid+cowpea-cowpea-berseem	255
2.	Cowpea-sorghum+cowpea-berseem	176
3.	N-B hybrid+guar-lucerne	253
4.	Maize+cowpea-maize+cowpea-oats	169
5.	N-B hybrid+cowpea-cowpea-berseem	176
6.	Sorghum+cowpea-berseem+sarson-sorghum+cowpea	169
	III. Eastern Region	
1.	Bajra+cowpea-maize+cowpea-oats	103
2.	Maize+cowpea-sorghum+cowpea-berseem+sarson	96
3.	Maize+cowpea-Deenanath grass-oats	131
4.	N-B hybrid alone	106
5.	Maize+cowpea-maize+cowpea-oats	85
	IV. Southern Region	
1.	N-B hybrid+luceme	225
2.	Sorghum+cowpea-maize+cowpea-maize+cowpea	110
3.	Guinea grass	135
4.	Congo grass and guinea grass under coconut tree	75

Model Plan for Fodder Production Round the Year

In the greater part of the country there are two lean seasons, April to June and November to December, in which the green fodder supply is scarce. Under these conditions cropping plans have been developed to ensure year round supply of green fodder. Overlapping cropping sequences are useful for continuous supply of fodder round the year. The staggered planting system will enable daily harvest of lush green fodder. One plan is given below in table-2 and it can be tailored before adoption in other areas to cater the local needs. This plan has been recommended for irrigated areas in north-west India (Punjab, Haryana, Delhi, Western Uttar Pradesh and Rajasthan).

The entire area may be divided into plots and these rotations can be adopted with the following considerations:

1. In order to maintain continuous flow of green fodder supply for the animals each of crops may be split sown twice or thrice with weekly or fortnightly intervals depending upon the needs.
2. The sowing regime can be rotated with different crop sequences so as to maintain soil productivity and to check incidence of pest and disease menace.

The surplus production, if any, during lush period may be conserved as silage or hay to cater the feeding requirements during lean periods.

Table 21.2 : Model Plan

Sl.No.	*Crop Sequence*	*Season*	*Duration (Days)*	*GFY ($t\ ha^{-1}$)*
1.	Sorghum-berseem-bajra			
	Sorghum	June-Sept.	95	41
	Berseem	Nov.-Feb.	176	87
	Bajra	March-June	92	75
2.	Maize+cowpea-berseem-bajra			
	Maize+cowpea	June-Sept.	77	40
	Berseem	Oct.-March	180	90
	Bajra	March-June	92	73
3.	N-B hybrid intercropped with berseem			
	N-B hybrid	Round the year	180	140
	Berseem	Oct.-March	176	63.6

GFY: Green fodder yield.

Forage Production during Lean Period

In India agriculture, most of the cultivated lands under irrigation are covered under double cropping system. Under this type of cropping there exist a scope of for growing short duration fodder crops during fellow period. *e.g.* wheat-maize/jowar/bajra rotation, a gap period exist between April-June to fill up this gap usefully and to insure better utilization of resources, fodder mixtures like maize cowpea bajra, cowpa or chari+cowpa can be produced without affecting the yield of grain crops of the rotation. Similarly, another gap period of relatively shorter duration exists during October-early November during this period short duration fodder like turnips and some *Brasica* species can be grown for forage. Cultivation of forage crops during this gap period provide around 30 t/ha of green fodder. While adopting the later scheme early maturing crop varieties of *kharif* season (*e.g.* maize for cobs) should be selected only with proper fertilizer and irrigation schedules to ensure early bulking of crops for the lean period find to eliminate the effect of late sowing in subsequent *rabi* crop. This plan is particularly useful for small farmers who have irrigation facilities but cannot divert land area exclusively for forage production. For big farms, *rabi* maize and legume with longer growth period like field bean serve the purpose well.

Forage Production in Drylands

Different kinds of grasses grow themselves in the naturally occurring pastures. Apart from this some species of grasses are grown in grassland areas entirely devoted to the culture of forage grasses and legumes in combination, purely under rainfed conditions. When grassland farming of pasture is practiced organic matter is renewed, soil erosion prevented, gully formation arrested, soil tilth improved. Sorghum, pearl millet, cowpea and cluster bean are the major grain cum fodder crops in most of the dry land areas. In semi-arid tropics with uncertain rainfall intercropping and mixed cropping are widely practiced to reduce the risk of complete crop failure.

Forage Production in Agroforestry Systems

Agro-forestry is the integration of woody species (tree/shrubs) along with arable crops/pasture grasses and legumes in the same piece of land. This is useful for optimising productivity, conserving plants, soil and nutrients and producing forage, timber and firewood on a sustainable basis. The forage availability in the traditional land use system is limited to rainy season only under rain fed farming, but in silvipastoral system round the year supply of forage is assured. Establishing suitable components of silvipastoral systems deserve consideration. Species for afforestation should have:

1. Deep root system
2. Ability to produce plants with high litter to recycle the nutrients
3. Should be fast growing and withstand frequent lopping
4. May have nitrogen fixing or mycorrhizal association to grow well on poor soils

A number of tree species (*Acacia sp., Ahlianthus sp., Albizia sp., Hardwickia* sp.) legumes (*Stylo, Desmodium, Clitoria*) and grasses (*Cenchrus, Dicahthium*) have been recommended. Such combinations conserve the soil and moisture and improve the productivity. The cultivation of improved range grasses like *Cenchrus ciliaris, Dichanthium annulatum, Lasiurus sindicus, Panicum antidotale* (depending on agro-ecological regions) and legumes like *Stylos, Siratro* and *Dolichos, Prosopis cineraria, Zizyphus* sp. and leguminous shrubs like *Leucaena, Sesbania, Glyricidia* as protein/energy source can mitigate the shortage of quality forage till end of winter season. Forage production potential of some crop rotation with grasses is given in Table 21.3.

Table 21.3: Forage Production with Perennial Grasses

Crop Rotation	*Green Fodder Yield (t ha^{-1})*
Hybrid napier-berseem/cowpea	250
Guinea grass - berseem/cowpea	200
Hybrid napier – subabool	235

Food-Fodder Production Systems

The small and marginal farmers with small land holding are not able to allocate any part of their land exclusively for forage production in main cropping season. Hence an integrated approach to food fodder production system aims at obtaining food as well as fodder concurrently from the same piece of land. In most of the irrigated areas depending upon the water availability double or triple cropping is prevalent. In such areas there is scope for growing short duration fodder crops during the fallow periods or in between two crops. In dry lands short season fodder crops like cowpea, guar etc., yielding 20-25 t ha^{-1} of green fodder within 45-50 days can be grown without affecting the main season crops. Barley can be grown as food cum fodder crop by adopting the simple practice of harvesting the crop at 55 days after sowing for fodder and managing the ratoon for grain purpose. In case of maize it is possible obtain both grain and fodder by sowing the crop at 30 cm spacing with 50 kg ha^{-1} seed rate and harvesting alternate rows at 50 days for fodder and allowing the remaining rows for grain. Intercropping of legume fodder crops with grain crops also help to produce good fodder.

Forage Conservation

It is essential to conserve surplus nutritious green forages at the appropriate time, with a view to tide over the problem of lean periods. The conserved forages would serve the important purpose of supplementing straw based rations of low nutritive value on which livestock usually subsist during the lean periods.

The methods of conservation generally practiced, are silage and hay making in India. Silage is the second best and cheapest cattle feed if preserved at the right stage of plant growth and made available for feeding in the lean periods. Silage making is not popular amongst Indian farmers. The farmers make hay from natural grasses when they are quite mature and poor in nutritive value. Modern technology also makes use of artificial dehydration of herbage for producing hay of high quality irrespective of weather conditions.

Silage Making

It is the wet preservation of green forages where sufficient acid is formed rapidly under anaerobic condition which acts as preservative. The most important substrate for formation of acid is water soluble carbohydrates. In the classical process of silage making, soluble carbohydrates in fodders are converted by lactic acid bacteria mainly into lactic acid and other organic acids. The resulting pickle-like product is acidic, ideally having a pH value around 4. At this pH, harmful bacteria *e.g.* those producing butyric acid is inhibited and the degradation of protein to ammonia and amines is prevented. The desired pH can usually be attained by proper compaction of the chaffed fodder, so as to ensure practically complete exclusion of air from the content of the silo. Exhaustive literature is available on type of silo, crops and stage suited for silage making, steps to be followed for making good quality silages, advantages and disadvantages of silage and feeding value of the silages. Recently. Following important points for silage making.

- ✩ Crops and plant material rich in soluble sugars such as maize, sorghum, oat, sugarcane tops and grasses are highly suitable for ensiling;
- ✩ The dry matter concentration of the forage at the time of ensiling should be around 25-35 per cent but higher if possible;
- ✩ Chaffing of the material for ensiling increases its compactness, thus eliminating the space to the maximum extent;
- ✩ The silo should be air tight after filling;
- ✩ Fermentation starts 'within hours after closing the silo', and accelerates over the next 2-3 days. It terminates after three weeks. Organic acids, primarily lactic and acetic acids, ethanol and gases such as CO_2, CH_4, NO_2 and NH_3 are produced during the fermentation process;
- ✩ Due to the production of acid, the pH of the biomass is reduced to a level below 4, resulting in the termination of all biological activities after which the material remains conserved under anaerobic condition. Under high dry matter silage (haylage) the pH remains around 5 or above.

Haylage

Due to high moisture content of silage, the dry matter intake by the livestock becomes a limiting factor. To overcome this drawback, high dry matter silage (called - haylage) is being advocated. The dry matter in haylage ranges between 45 to 65 per cent . The technique of haylage making is the same as that of silage making. However, utmost care is given in pressing because of the difficulty in excluding air from the material.

Hay

Green forage when dried to about 80-85 per cent dry matter, containing most of its nutrients including carotene, is called hay. The basic principle of hay making is to reduce the moisture concentration in the green forages sufficiently as to permit their storage without spoilage or further nutrient losses. The moisture concentration in hay must be less than 15 per cent at storage time. However; crops with thin stems and many leaves are better suited for hay making as they dry faster than those having thick and pithy stems and small leaves. Natural grasses, berseem, lucerne, cowpea, guar and oat are highly suitable crops for hay making.

Artificial Dehydration

Artificial dehydration of grasses and leguminous forage is practiced widely in the advanced countries. This is very suitable for regions, where climatic conditions are unfavorable for hay making. Another advantage of this method is that the nutritive value of the original crop is almost wholly retained in the final product. Due to high fuel energy cost, this method has limited applicability under Indian condition except for large dairy farms. The best feeding strategy for lean period may be to feed livestock about 1/3rd dry matter through crop residue, 1/3 rd from green fodder and remaining 1/3rd through conserved forage with strategic supplementation of deficit nutrient.

The production of forage crops can be improved through intensive forage production system in irrigated areas and better management practices in dry land areas. However, supply of green fodder throughout the year is not possible due to climatic conditions, fodder conservation can play important role in filling the gap between demand and supply of green fodder during lean period.

Chapter 22

Strategies for Improving Dairy Herd Performance

S.K. Kaushish

Central Arid Zone Research Institute,
Jodhpur – 342 003, Rajasthan

The country has large number of genetically diverse livestock resources; we have made large research and development efforts in improving their production and productivity during the last two decades. This is reflected in an increase in milk production from 22.5 million tonnes to 78.2 million tonnes, egg production from 5.3 million to 98 million during the last 20 years. However, considering the current and projected demand of livestock products, the rate of increase in products will have to be substantially improved.

In the seventh five year plan the annual rate of progress was 6.5 per cent for milk and 8.0 per cent for eggs and which increased to 7.2 per cent and 8.5 per cent, respectively, during Five Year Plan. The largest contribution from agriculture sector to GNP is that of rice, milk is second and wheat is at the third place. The acceleration in the rate of livestock development is possible with the application of techniques in Animal Reproduction and health. The emergence of newer bio-technologies like embryo transfer, sexing, clonning and genetical engineering, could make it much faster. At present, these technologies are costlier and still under experimentation in laboratories, however, further research will make them economically feasible in near future.

Genetic Improvement

Extensive programmes for upgradation of genetic potential of indigenous livestock through crossbreeding with superior exotic breeds, grading with superior indigenous breeds and replacement of the indigenous breeds by the exotic superior breeds has been in progress. This has been supported by research and development programmes on cyro-preservation of semen and artificial insemination. Superior strains of dairy cattle *viz.*, Karan Swiss and Karan Frise have been developed which yield 3200 and 3500 L of milk per lactation.

Breeding Policy and its Implementation

The country has large and genetically diverse livestock population of different livestock species. Inspite, there exists a broad gap in breeding policy. The details with respect to the improved breed and the breeding programme for each agro-ecological and socio-economic situation have not been worked out and implemented. Further in case of dairy cattle and buffaloes, genetic improvement, in milk production has been limited due to problems in selection of sires to be used for breeding on the basis of their genetic merit. It is, therefore, necessary that detailed breeding policy with respect to different livestock species keeping in view the ecological conditions, feed resources and demand of products, with respect to improver breed and the breeding methodology is developed. In addition facilities should be developed for selection of breeding males on the basis of their genetic merits. This will require creating new farms, expanding the existing farms, linking the existing organised breeding farms maintaining the same breed/ breed cross and performance recording of pedigreed progenies to allow progeny testing of a large number of bulls and their subsequent selection. Alternatively, we should develop systems of field recording of performance data of farmers' animals and their utilisation in progeny testing and subsequent selection of bulls. The former approach would have the limitations of the quantity of superior germplasm available relative to that required. The latter would have the limitations of organisation and cooperation of the farmers in allowing recording of performance. However, with the multiple ovulation embryo transfer technology (MOET) available, it may be desirable to obtain as many male progenies from a few elite families and use them extensively through artificial insemination using frozen semen, which may be more expeditious, though not technically and effectively as sound.

Improvement of Non-Descript and Low Productive Zebu Cows by Crossbreeding with Exotic Breeds

Majority of cattle of our country are non descript and yield very little. Quantity of milk. In milk shed areas around cities and industrial towns, where, good market for milk is available and where good resources of feed and fodder exist, crossbreeding of these low milk producing cows with exotic breeds should be taken up. The choice of the exotic breed in the plains may be Holistein. The investigation conducted so far indicated that the optimum level of exotic inheritance is 50 per cent.

Table 22.1

Strain	*Exotic Inheritance*	*First Lactation Milk Production (305 days)*	*Over All Lactation Average (Kg.)*
Holstein X Tharparkar	50 per cent	3457	3755
Brown Swiss X Tharparkar	50 per cent	2816	2953
Jersey X Tharparkar	50 per cent	2659	2804

Improvement of Indigenous Non-descript Cows by Grading Up

In areas, where, good renumerative market for milk is not available or in areas, where good resources of feed and fodder are not available and for farmers who do not like to go in for crossbreeding, it would be desirable to improve the indigenous non descript local cows by grading up to better indigenous breeds known for fairly good level of milk production. The indigenous breeds with fairly high level of milk production are Sahiwal, Tharparkar, Rathi, Kankrej, Red Sindhi, Gir and Ongole. Farmers' herds in Gujarat could be graded up to Gir and Sahiwal, the farmers' herds ino Sourthen, Northern and Western Rajasthan could be graded upto Kankrej, Rathi and Tharparkar, respectively. Punjab, Haryana, Uttar Pradesh and Bihar cattle could be graded upto Haryana.

Improvement of Indigenous Low Producing Cows by Selective Breeding

Only about 15 - 20 per cent of cows belong to well define breeds. These breeds have been evolved in various agro-climatic and ecological regions of the country according to regional needs. These animals have survived under stressful conditions of the tropical environment, limited quantity of feed and fodder availability etc. The farmers have over centuries tried to select these breeds by and large for draft. Only few breeds like Tharparkar, Rathi, Kankrej and Sahiwal were selected for milk production. More herds of known indigenous breeds mentioned above should be established. The male calves with high pedigree index for milk production should be raised and tested for semen quality. The semen of these bulls should be used in farmers' cows of the same breed.

The areas of the country, where the indigenous breedings need to be improved by selective breeding are:

- ☆ Gujarat for Gir and Kankrej
- ☆ Rajasthan for Rathi, Tharparkar and Nagori
- ☆ Punjab, Haryana and Western V.P. for Haryana
- ☆ Karnataka for Hallikar, Deoni and Amrit Mahal .
- ☆ Tamil Nadu for Kangayam
- ☆ Andhra Pradesh for Ongole

Genetic Engineering Other Biotechnologies in Genetic Improvement and Genetic Resource Conservation

MOET coupled with sexing and cloning can help in faster genetic progress through selection by increasing efficiency of selection compared to conventional schemes based on progeny testing of sires. Further methods for genetic engineering especially introducing of superior extraneous genes through micromanipulation in fertilized embryo and development of transgenic animals have more recently been developed. Examples of such transgenic animals are available from a number of countries - one being of fast growing sheep in Australia. Method for genetic resource conversation through cryo preservation of semen and embryo fragments of chromosomes and genes, the latter in bacterial or other cellular systems are now available and should be utilised to develop gene banks. Methods for selection of livestock based on restricted fragment length polymorphism (RFLP) could be utilised in selection of superior genotypes. Similarly gene marker characters of biochemical, immunological and cytological in nature could be utilised in genetic differentiation of breeds and strains, studying their evolutionary history and possibly in selection for better production and adaptation.

Improving Reproductive Performance

Maximising animal reproduction can tremendously help in increasing production of animal products. Major emphasis in research should be in studying the hormonal profiles in important breeds of livestock under various physical, environmental and nutritional conditions and develop managemental practices, which will maximise reproductive performance. Administration of bovine growth hormone to dairy cattle can increase milk production by 20 per cent without serious effect on health and is nutritionally efficient. Research and development efforts in improving success rate of A I and for more effective long term preservation of semen especially cryo-preservation of ram and buck semen are required for enabling large scale crossbreeding programmes involving exotic breeds being undertaken for improvement of these two species.

Reducing the Age at First Calving and Intercalving Interval by Improved Managemental Practices

Bos indicus species of cattle are known as low milk yielder with poor reproductive efficiency in comparison to *Bos taurus* (Exotic cattle). By following improved managemental practices it is possible to increase milk production from Tharparkar breed of cattle. Normally the heifers of this breed first calve at the age of 4 to 4.5 years and the intercalving period is 1.5–2 years under field conditions. The scientists of the division of the Animals Sciences and Forage Production of CAZRI, Jodhpur have succeeded in reducing the age at first calving to an average of 35 months and the intercalving interval to less than 1 year (Kaushish, *et al.*, 1998, Patel *et al.*, 2000). The Tharparkar cattle maintained at this institute have attained the above ideal targets, a rare achievement for the indigenous breed in the country. The package of practices for this fertility optimization are as follows.

1. *Balanced Feeding*: Providing a balanced concentrate mixture round the year, irrespective of the lactation or dry stage of the animals to remove nutritional deficiencies, particularly hypoproteinaemia is considered as the most effective step. The balanced concentrate mixture consisting of pelleted concentrate feed (65 per cent), Tumba seed cake (25 per cent) and Oil cake (10 per cent). The quantity of the above concentrate mixture given to heifers (3 months to 1 year), 1 year till confirmed conception, dry cows and pregnant heifers/cows was 0.5-1, 2 and 2.5 kg/day, respectively. Alongwith concentrate feed vitamin-mineral mixture is fed from January to June.
2. *Feeding of ecbolic mixture to post parturient animals*: The cows immediately after calving are fed ecbolic powder @ 100 gm per head per day for 10 days to hasten the uterine involution.
3. *Timely treatment of gynaecological disorders*: The post parturient animals are gynaecologically examined 2 months after calving to rule out the persistent corpus leutium and uterine disorders. The repeat breeders are also given timely specific treatment depending on the condition.

Normally the farmers do not give any concentrate to their heifers till they calve. By feeding and following above practices the age of maturity is reduced by one and a half year and intercalving period by 6-10 months, thus there is a net increase in income by about Rs. 75,0001- to the farmers from one cow.

Economical Feeding of Cattle Utilizing Non-Conventional Feed Resource–Tumba Seed Cake

It has been observed that the cattle feed having 25 per cent Tumba: *(Citrullus colocynthis)* seed cake provided to Tharparkar cattle has no adverse effect on both the productive and reproductive performance. Tumba seed cake is locally available in abundance (having 18-20 per cent CP) and is a cheaper source of protein and is normally used as a fuel for furnances and this wasted. This technology results in a net monetary saving of around. 18-20 per cent without impairing the productivity. This combination can be continued without any ill effect and lot of monetary saving is achievable and there is a savin of conventional feed also (Mathur *et al.*, 1997).

For a unit of 4 cattle there is a net saving of Rs. 20 per day which comes to Rs. 7300 for one year and to Rs. 42800 during a productive life span of six years.

Physico-chemical Method for Detanning of *Prosopis cineraria* Leaves to Improve their Nutritive Value

Khejri, *(Prosopis cineraria)*, is widely distributed in arid region of Western Rajasthan. Its leaves are consumed by the livestock as browse, branches are lopped and offered to range foraging animals and the trees are systematically lopped and the leaves are collected, stored to offer the corraled animals. *P. cineraria* leaves contain appreciable quantities of protein (15 per cent) but experiments revealed that their digestible crude protein is very low (5-5.5 per cent). This low crude protein digestibility is attributed to their high tannin content. Attempts have been made to develop an appropriate technique to remove the nutritional constraint. Suggested method involves soaking of the leaves in 1 per cent aqueous sodium carbonate solution for six hours during the day. The supernatant is removed and leaves after air-drying can be offered to the animals. By this technique as high as 94 per cent of tannins present in the *P. cineraria* leaves can be removed. (Bohra and Kaushish).

Multi-nutrient Blocks for Supplementary Feeding

Desert livestock mainly thrive on agricultural by-products, stubble and ripened grasses. These feeding stuffs are highly fibrous and nutritionally very poor. Due to their low density, they occupy enormous space. As such an attempt has been made to densify, by means of compaction, after fortification with nutritionally rich ingredients like molasses, urea, mineral mixture, dolomite, common salt, alongwith a suitable binder to locally available roughages like chaffed *C. ciliaris* and *L. sindicus* straw, air dried and powdered *Ardu* leaves, wheat bran or mixture of these feeds. After thorough mixing, all these ingredients are compressed in a wooden mould to make a block. Feeding of such grain-free, roughage-based multi-nutrient block is likely to save considerable quantity of grains normally consumed by human beings. The block supplemented animals have shown better growth as compared to control animals. These blocks can safely be stored for a long time and due to its compactness, it is easy to transport them in bulk quantities, especially, during the droughts (Anon., 1999).

Cultivated Pasture

The perennial monoculture grass *(Cenchrus ciliaris)* pasture in semi arid region has biomass yield of 2.0 T/ha which is sufficient to maintain 5 small ruminants or 1 ACU with followers round the year. The lambs born to the ewes maintained exclusively on Cenchrus pasture had birth, weaning, 6 and 9 months weight of. 3.2, 13.9, 20.6 and 23.9 kg, respectively with empty live weight dressing yield of 54.3 per cerit. Similar studies conducted on cultivated sewan *(Lasiuus sindicus)* pasture in arid location at stocking density of 5 sheep/ha indicated that the pasture intake was insufficient to maintain them from January to June. They required supplementation during lean period. Nutrient intake and utilization in sheep for supporting desired growth rate in lambs and to sustain adults stock in critical physiological stages, addition of pasture legumes *viz.*, Dolichos and Serato or intercropping with Cowpea has dafine advantage.

Sillvipasture

Establishment of silvipasture improves quality of nutrient available ensuing its supply round the year. Experiment conducted at CSWRI has indicated that the hoggets are able to achieve 30 kg at yearling age, whereas, under grazing on natural pastures with some supplementation with grass and cultivated fodder hay it was 25.0 kg. Similar studies conducted on ewes in advanced stage of gestation and lactation indicated that pasture intake was not adequate for them requiring supplementation to harvest desirable production traits. Another study with weaner lambs and kids on three and two tier silvipasture indicated that lamb and kids were able to achieve 22.0 kg body weight at 6 months of age without concentrate supplementation.

Grazing with Supplementation

The slaughter weight is far lower and the age at which target weight is achieved is far higher than expected. The system requires input of concentrate which although is cost effective, yet has not been adopted due to socio-economic condition of the farmers. The work done in the country on grazing with supplementation has indicated that in addition to free grazing supplementation of limited quantity of concentrate amounting to 1.5 to 2.0 per cent of body weight will provide marketable finishing weight of 25 and 30 kg at 6 and 9 months of age, respectively. The lambs under sole grazing on Cenchrus pasture with concentrate supplementation @ 1.5 per cent of their body weight achieved 26.7 kg at 6 months of age with about 100 g ADG.

Animal Health Programme

The animal health research and development has led to better understanding of disease problems in the country and development of effective diagnostic and prophylactic measures against them. A number of new and more effective vaccines has been developed. The country is self sufficient in diagnostic reagents and vaccines against most of the important livestock and poultry diseases.

These developments, as indicated earlier, have resulted in increase in production of major livestock products. However, considering the present and project future demand, the rate of increase will have to be substantially improved. A number of limitations in our technologies and their applications exist. Further a number of newer biotechnologies are now available which must be applied for increasing the rate of improvement in livestock production.

Improving Disease Forecasting, Control and Eradication Measures

Although indigenous breeds of livestock have developed relatively better resistance to tropical diseases through natural selection, with the effect to increase their production especially through the introduction of temperate inheritance in them, the importance of even the prevalent diseases and the new emerging and exotic diseases is being realised. It is necessary to develop more effective disease surveillance, monitoring and forecasting systems to allow more effective control and eradication measures to be instituted. It will also be necessary to prepare the country to meet the challenges of emerging and exotic diseases and for the purpose high security animals' disease facilities will be required. India is perhaps the first in the developing countries to consider setting up such facilities. Newer biotechnologies especially related to DNA recumbinance are available for developing more effective and cheaper immuno-diagnostic and immuno-prophylactic measures against important diseases. The greatest importance of such technologies is in developing diagnostic tests which can be utilising sophisticated instrumentation. The cloned protein vaccines have also less problems of storage and transportation and are more after and cheaper. Blood protozoan diseases transmitted by ticks are gaining importance and there is a need for developing effective prophylaxis against them. Such work is in progress and some satisfactory lead has been obtained.

अध्याय 23

गौवंश, दुग्ध एवं दुग्ध उत्पादो की वैदिक अवधारणा

रामचन्द्र पाण्डेय

पूर्व संकाय- प्रमुख एवं विभागाध्यक्ष
ज्योतिष विभाग, काशी हिन्दू विश्वविद्यालय, वाराणसी

वैदिक साहित्य में गौ का अत्यन्त आदरणीय स्थान है। गौ शब्द का प्रयोग पृथ्वी और सूर्य रश्मियों के लिए भी किया गया है। समस्त जैविक सृष्टि का आधार सूर्य और पृथ्वी ही है। सभी प्रकार के जीव–जन्तु एवं वनस्पतियां सभी पृथ्वी पर उत्पन्न होते हैं तथा सूर्य रश्मि के सहयोग से पृथ्वी द्वारा उनका भरण पोषण होता है। इन दोनों प्रधान स्टृष्टि स्रोतों के प्रतिनिधि के रूप में गौ का माना जाता है। जब–जब पृथ्वी पर सन्तुलन बिगड़ता है भौतिक उत्पातों की वृद्धि होती है जीवन असुरक्षित हो जाता है तब–तब पृथ्वी गौ का रूप धारण कर सृष्टिकर्त्ता के समक्ष उपस्थित हैं। ऐसे अनेक अवसरों के उल्लेख पुराणों में उपलब्ध हैं। जिनसे गौओं की श्रेष्ठता परिलक्षित होती है। एक वैदिक सन्दर्भ से ज्ञात होता है कि पहले गौंओं की सृष्टि नहीं थी जब उनकी महती आवश्यकता समझी ई तब ऋषियों नें "अग्ड़िश" ऋषि से गौ प्राप्ति के लिए अनुरोध किया। अग्ड़िश ने घोर तपस्या कर गौओं को प्राप्त किया।[1]

उत्पत्ति के बाद स्टृष्टिकर्त्ता ब्रम्हा ने गौओं को आशीर्वाद के साथ आदेश दिया कि–"हे धनवान गो समूह अपने दोहन स्थान में एवं दोहन के विचरण स्थान में सुख पूर्वक रमण करो। इस लोक के सभी क्षेत्रों में, अपने पालक के घर में तेज के साथ निवास करो अन्यत्र न जाओ।"[2]

वैदिक काल में यज्ञ प्रमुख कर्म थें। यज्ञों के प्रभाव से प्राकृतिक सन्तुलन बना रहता है तथा सुभिक्ष एवं सुवृष्टि से समाज सुखी और सम्पन्न रहता है । इन सभी यज्ञों में हवि गव्य से विशेष रूप से दूध और घृत से तैयार की जाती थी। गोमय के कण्डों पर अग्नि का आवाहन होता था तथा गोमय के उपलों पर ही 'पुरोडाश' पकाया जाता था। अन्य गव्य यथा 'गोमुत्र', स्थान एवं शोधन हेतु प्रयोग होता था। इस प्रकार यज्ञ में सभी गव्यों (दूध–दधि–घृत–गोमय–गोमूत्र) का प्रयोग होता रहा है। गौओं के रखने (बाँधने) के स्थान को गोष्ठ कहा जाता है। जिस क्षेत्र में गोचारण होता था उसे भी गोष्ठ ही कहा जाता था। जहाँ गायों का दोहन होता था उस स्थान को 'योगि' कहा जाता था। जैसा कि उक्त मन्त्र मे उल्लिखित है।

गायों के रक्षक देवता का नाम 'पूषा' है। इन्हें 'आनष्ट पशु' भी कहा जाता है। 'पूषा' सूर्य के अनेक नामों में से एक है। ऋग्वेद में इसका इस प्रकार उल्लेख है –'अवनष्ट पशु पूषा लोको में विचरण करने वाली गौओं (पशुओं) की रक्षा करें।"[3]

गौओं की संज्ञा एंव वर्ण (रंग) प्रायः चार वर्ण (रंग) की गौओं का उल्लेख मिलता है। 1. रोहित (लाल), 2. शुक्र (सफेद), 3. पृश्नि (चित्रित या चितकबरी) कृष्णा (काली)। गौओं को उनके वर्णों से भी जाना जाता था। अर्थात् वर्ण ही उनके नाम भी हो जाते थें यथा रोहित, शुक्र, पृश्नि, कृष्णा आदि।

वर्ण कें अतिरिक्त इनके वर्गीकरण का आधार इनकी विभिन्न अवस्थयें भी थीं यथा– सफेद गाय की संया 'कर्कि, गर्भिणी गाय की 'गृष्टि', दूध देने वाली की 'धेना या धेनु', बन्ध्या (जो प्रजनन योग्य नहीं होती) को 'स्तरी' या धेनुष्टर', काकबन्ध्या (जिसे केवल एक ही बच्च हो अर्थात एक सन्तान के बाद पुनः प्रजनन न हो) की 'सूतवत्सा' जिसका गर्भपात हो जाता हो उसकी 'वेहत्' संज्ञा होती है। जिस गाय का बछड़ा जन्म लेकर नष्ट हो जाय तथा

बछड़ेकभी देखकर दूध देती हो उसे ''बान्था या निवान्या'' कहा जाता था। इस प्रकार प्रकृति के अनुसार सात वर्गो में गायों को परिगणित किया गया है। इतना ही नहीं समूह में से विशिष्ट गायों की पहचान के लिए उनके शरीर पर कुछ चित्र अंकित किये जाते थें। उन्हीं चित्रों के नाम से उन गायों का भी नाम करण होता था। यथा–जिस गाय के शरीर पर वंशी का चिह्न होता था उसे कर्करिण्य संज्ञा दी जाती थी।जिसके शरीर पर हंयिा का चिह्न हो तो स्थूण करिझय, तथा आठ अंक का चिह्न हो तो उसे अष्टकण्यि या अष्टकर्णि कहा जाता था। अपनी–अपनी गायों की पहचान के लिए प्रायः सभी लोग उनके शरीरर पर कोई न कोई चिह्न लगा देते थें। उन्ही चिन्हों के ना`स गौ को भी पुकारते थें।

कुछ इस प्रकार की संज्ञाओं का भी उल्लेख वेदों में मिलता है जिनसे उनकी (गाय–बछड़ों क्री) आयु का ज्ञान होता है। यथा दुधमुहें बच्चे की 'धारण' कहा जाता था। इसी प्रकार ड़ेढ वर्ष की बछड़ी को 'यवि' तथा बछड़े को 'यैवि' कहा जाता था। इस प्रकार के नामों (संज्ञाओं) से बछड़ों की आयु का भी ज्ञान हो जाता था। आयुबोाक संज्ञाओं की वूची निम्नलिखित है–

आयु	*गाय (बछड़ी)*	*बैल (बछड़ा)*
सद्यःजात(दुधमुहाँ)	धारण	धरण
ड़ेढ वर्ष	यवि	यैवि
दो वर्ष	दित्यौही	दित्वाह
ढाई वर्ष	प०यावी	प०यावी
तीन वर्ष	त्रिवत्सा	त्रिवत्स
तीन वर्ष छः माह	तुर्यौही	तूयवाह
चार वर्ष	प्रष्टौही	प्रष्ठवाह
युवा बैल को	–	ऋषम् वृष्य
गाड़ी खीचने वाले बैल को	–	अनड्वान्
बचिया बैल को	–	महाविरष्ट

गोदोहन

दिन में तीन बार गोदोहन होता था। तानों संध्या काल कोदोहन काल होता था। प्रातः कालीन दोहन को प्रातर्दोह, मध्याह्न काल के दोहन को संगव दोह तथा सांयकालीन दोहन को सांय दोह कहा जाता था।[4]

'संगव' दोह का शाब्दिक अर्थ मध्याह्न काल का सूचक नहीं है। अतः सम्भव है संगव काल में ही दोहन होता रहा 'संगव' काल मध्याह्न और प्रातः काल के बीच का काल होता है।

सामाजिक मान्यता

गाय को सम्पत्ति का मानक माना जाता था। इसलिए आज भी इसे 'गोधन' कहा जाता है। एक बारत बाम देव ऋषि ने क्रोध में आकर पूछा कि 'वह कौन है जो मेरे इन्द्र को दस गायों में खरीदना चाहता है।5 यह तो हजार या दस हजार गायों से भी नहीं खरीदा जा सकता।'' शतपथ ब्राम्हण भी गायों को सम्पत्ति का सूचक माना है। दक्षिणा के रूप में द्रव्य के स्थान पर गायों को देने का निर्देश देता है।[6]

ऋग्वेद तो गायों को ध नही न ही अपितु सर्वस्व ही मानता है कहा गया है –गाही सूर्य है, गाय ही इन्दू है गाय ही सर्वस्व है इत्यादि।[7] इस प्रकार गायों की महत्ता प्रतिपादित करते हुये गौ की प्रार्थना की गई है–''गायें मेरे आगे हों, मेरी पीछें हो, गाय ही मेरा हृदय हो, हम सदवें के बीच में निवास करें।[8]

अभिप्राय यही कि हमारे लिए सभी प्रकार से उपयोगी, सम्पत्ति सूचक तथा गव्यों के कारण यज्ञ आदि में अत्यन्त उपयोगी गौयें हमारे पास सदैव प्रचूर मात्रा में रहें।

प्राचीन काल में स्पष्ट मान्यता थी कि 'यज्ञ' लोक कल्याण हेतु अत्यन्त आवश्यक है। यज्ञ से ही वांदल बनते हैं वादलों से वृष्टि होती है तथा वृष्टि से ही अन्न उत्पन्न होते हैं।[9] आज भी विविध प्रयोगो द्वारा यज्ञ की सार्थकता सिद्ध हो रही है। तथा अत्तोत्पादन में भी यज्ञ की अनुकूल भूमिका सिद्ध हो रही है।

अन्न की आवश्यकता

महर्षि पराशर ने लिख है कि अन्न ही प्राण है अन्न ही बल है, अन्न ही सभी कार्यो का साधक है। न केवल मनुष्य ही अपितु देवता और असुर आदि सभी का जीवन अन्न पर ही आश्रित है।[10]

कष्ठ में, कानों में तथा हाथें में यदि स्वर्ण आभूषण हो तो भी अन्न के अभाव में उपवास ही होगा।[11] इसलिए कृषक को कृषि हेतु सावधान तथा स्वयं समर्पित रहना चाहिये। कृषि पाराशर में कहा गया है कि

पितुरन्तः पुरं दद्यान् मातुर्दद्यान्महानरूम।

गोषु चात्मरुमं दद्यात्) स्वयमेव कृषि व्रजेय्।।

अर्थात्–पिता को आनन्दपुर (गृहस्थी सुरक्षा में लगा दें) माता को रसोई (किचन) में लगा दें। अपने समान व्यक्ति को गौंओं की देखभाल में लगा दे तथा कृषि के लिए स्वयं ही जाना चाहिये। और भी कहा है–

गो हितः क्षेत्रगामी च कालज्ञों बीज तत्परः।

वितन्द्रः सर्वशस्यढत्रचः कृषको नावसिदति। कृ.पा.

गौवों का हित चाहने वाला,पन अपने खेतो परनाने वाला कृषि कर्म योग्य काल जानने वाला, बीज विशेषज्ञ कृषक कभी भी कष्ट नहीं पाता।

गोशाला

जिसकी गोशाला मजबूत गोवर से रहित सदैव साफ सुथरी होती है उसके गाय बैल बिना बौष्टिक आहार के भी सदैव पुष्ट स्वच्छ एवं प्रसन्न रहते हैं।[12]

इस प्रकार गोशाला, गोमय छूट, खाद आदि का विस्तृत विवेचन किया गया है। उनके विस्तार में न जा जाते हुये अब गव्यों के विवेचन पर प्रकाश डालते हैं।

गव्य

गाय से प्राप्त होने वाले पदार्थो को गव्य कहते हैं। ये पाँच प्रकार के होते हैं।[13]

1.गोमूत्र; 2. गोमय; 3. दुग्ध; 4. दधि; 5. घृत।

इनमें कुशोपक मिला देने से सुरक्षित एवं शुद्ध प प०च गव्य हो जाता है। प०च गव्य मे वरूण आदि देवों को निवास माना जता है इसलिए इसके पान से आत्मशुद्धि तथा भूमि पर छिड़कने से भूमि की शुद्धि होती है। इसके निर्माण की विधि शास्त्रों में निम्नकम से दी गई है:

गोमूत्रः 1 पल

गोमयः अंगुष्ठ की मोटाई का आधा

दूधः 7 पल

दहीः 3 पल

घृतः 1 पल

कुशादकः पल

{पलमात्रं तु गोमूत्र अंगुष्ठार्ध तु गोमयम।

क्षीर सप्रयलं ग्राह्रम दधि त्रिपल मीरितम्।।

सर्पिस्त्वेक पलं देय मुदकं पल मात्रकम।।}

प०चामृत

गाय का दूध, दधि, घृत, मधु और शर्करा। इनके मिश्रण को प०चमृत कहा गया है। प०चगव्य और प०चमृत दोनो ही आत्मशुद्धि के पदार्थ होने के साथ–साथ यज्ञीय प्रमुख पदार्थ भी हैं। गाय के दूध में बने पुरोडास को दूध, दही और घृत से मुक्त कर देने पर 'सान्नाय' नामक हविष्य बनता है। इसी हविष्य से दर्श और पौर्णमास यज्ञों मे आहुतियाँ दी जाती हैं।

प०चगव्य और प०चमृत से स्नान करने तथा इनका पान करने से वाह्य–आन्तरिक और आध्यात्मिक शुद्धि साथ–साथ शरीरिक बल एवं सौन्दर्य की वृद्धि होती है।

प०चगव्यों का उपयोग

1. घृत दीर्घायु की कामना से अग्नि में घृत की लक्षसंख्यक आहुति दी जाती है।[14]

2. घृत के सेवन से बल और आज की वृद्धि होती है। साथ ही व्रण आदि के उपचार में भी घी का प्रयोग होता है। अन्धक–शिव युद्ध में अन्धक ने क्षत विक्षत हो जाने पर शिव का शिग रू॰ धारण कर पार्वती के पास जाता है। उस समय पार्वती ने अपनी दासियों से कहा– पुराना घी, वीजिका, ण और दधि लाओं मैं स्वयं शिव के ब्रणों का उपचार करुँगीं।[15]

घृत की परिभाषा और लक्षण

'पस्वनवतीतं घृतम' अर्थात् पके हुये मक्खन को घृत (घी) कहते हैं। घृत स्वाद में मधुर होता है। यह बुद्धि, शुक, ओज, स्मूति और कफ को बढ़ाने वाला तथा वात–पित्त–विष उन्माद, शोक और दरिद्रय को दूर करने वाला है।[16]

पुराने घृत का लक्षणे

दस वर्ष पुराना घृत उग्रगन्ध मुक्त तथा लाक्षारस के समान होता है। इसे पुराण घृत कहते हैं। दशवर्ष से अधिक पुराना होने पर उसे 'प्रपुराणघृत' कहते हैं।

कॉस्य पात्र में दश दिन भी रखा हुआ घृत विषाक्त हो जाता है अतः वह न्याज्य होता है।

नवनीत

नवनीत दो प्रकार का होता है 1. दधि से तत्काल निकाला हुआ नवनीत (मक्खन) 2. दुग्ध से निकाला हुआ नवनीत।

इन दोनों के गुण धर्मों में भी सामान्य अन्तर है।दधि से उत्पन्न नवनीत बालक और वृद्धों के लिए प्रश्नत्त है यह बलवर्धक होता है।[17]

दुग्ध से उत्पन्न नवनीत नेत्र ज्योति को बढ़ाने वाला तथा रक्तदोष और पित्तदोष को नष्ट करने वाला होता है।[18]

दधि

"क्षीरोत्तरावस्था भावः दधिः' अर्थात दुग्ध की अन्तिम अवस्था का अभाव हो जाना दधि कहलाता है। अर्थात जब दुग्ध का स्वरूप पूर्णरूपेण बदल जाता है उसे दधि कहते हैं। यह वात वृद्धि करने वाला तथा पिपासा को शान्त करने वाला होता है।

रात्रि में दधि सेवन का निषेध किया गया है।[19] यदि खाना ही हो तो उसमें शर्करा, मधु, मूँग तथा आवला मिलाकर खाना चाहिये।

तक्र (मट्ठा)– तीन प्रकार का होता है।[20]

1. दधि का घोल जिससे सम्पूर्ण घृत निकाल लिया गया हो।
2. दधि का घोल जिससे आधा घृत निकाल लिया गया हो।
3. दधि का घोल जिससे घृत न निकाला गया हो।

अन्य मत से तक्र (मट्ठा) पाँच प्रकार का होता है। 1. घोल, 2. मथित, 3. तक्र, 4. उद्स्चित तथा 5. छच्छिका।[21] नके लक्षण क्रमशः इस प्रकार हैं–

1. घोल– जब रहित दधि का घोल।
2. मथित– दधि को अच्छी तरह 'मथ' कर तैयार किया गया जल रहित घोल।
3. तक्र– दधि और उसकी मात्रा का वे जल।
4. उद्स्चित्– आधा दही–आधा जल मिलाकर मन्थन द्वारा तैयार घोल।
5. छच्छिका–दधि की मात्रा से अधिक जल डाल कर मन्थन किया गया घृत रहित घोल।

दुग्ध(क्षीर)–दुग्ध को सम्पूर्ण भोजन माना जाता है क्योंकि इसमें शरीर पोषक लगभग सभी तत्व पाये जाते हैं। दुग्धपान से आयु की वृद्धि होती है। इसलिए जन्मदिन पर बच्चों को दुग्ध, गुड़ और तिल पिलाया जाता है।[22]

साथ ही क्षीर से बना हुआ भोज्य पदार्थ पायस (खीर) का हवन करने[23] तथा इसका सेवन करने से बल–ओज–नैरूज्य की प्राप्ति होती है।[24]

क्षीर के भेद

दूध (क्षीर) दो प्रकार का होता है। 1. पेयूष, 2. मोरट। प्रसवकाल से सात दिनों तक का दूध पेयूष कहा जाता है। सात दिनों के बाद का दूध 'मोरट' कहलाता है।[25]

गोमूत्र

शरीर की आम्यन्तर एवं वाह्य शृद्धि करने वाला द्रव्य है किन्तु आयुर्वेद इसके अन्य गुणो का उल्लेख करते हुए इसे औषधि के रूप में वर्णित किया है इसके गुण–धर्म का विवेचन करते हुए लिख है–

'गोमूत्र कटु तीक्ष्णोष्णं सक्षारत्वान्न वातलाम्।

बह्वगि दीपनं मेध्यं पित्लं कफ वातजित्।

शूलगुल्मोदरामाह विरेका स्थपनादिषु।

मूत्र प्रयोग साध्येषु गव्यं मूत्र प्रयोजयेत्।।

यहाँ मूत्र के गुण और लक्षण को बताते हुए यह निर्देश दिया गया है कि जब भी मूत्र चिकित्सा देनी हो तो केवल गोमूत्र का ही प्रयोग करें।

गोमय

यज्ञाद्धि का आधर 'कण्डिका' के रू॰ मे प्रयोग में आता है। कण्डिका भस्म आरोग्य कारक तथा कृमिनाशक कहा गया है। इसीलिए गोमय का प्रयोग प॰चगव्य में शरीर शोधन हेतु भी किया जाता है। और गों से प्राप्त गव्यों पर विचार करने से यह सिद्ध होता है कि गौ पालन और गो सेवा निश्चय ही मानव मात्र के कल्याण के लिए है। अष्टाग्ड् हृदय की यह उक्ति "अत्र गव्यं तुजीवनीयं रसामयनम्" सर्वथा सत्य है। पुराणें में अनेक कथा ये हैं जो गो सेवा के महत्व को प्रतिपादित करती है। राजा दिलीप की गोसवा सव्रविदित ही है। आज भी गो सेवा दुःख– दारिद्रय और व्याधियों से छुटकारा दिलाकर मनुष्य को दीर्धायुष्य प्रदान करने में सक्षम है।

शार्

सन्दर्भग्रन्थ

1. ''या आरिससापसेह चक्रस्ताम्यः महि शर्म यच्छ'' ऋग्वेद 10 |169 |2

2. ''रेवती रमध्वमस्मिन् योनावस्नि् गोष्ठे स्मिंल्लोके स्त्रिन्क्षये इहैव स्त मापगाते ।'' शु. म. सं. 3.21

3. ''पूषा त्वेतश्चाव्यतु प्रविद्वानवनष्टपशु भुर्वनस्य गोपाः |ऋ 10 |17 |3

4. तै. सं. 7 |5 |3 |1

5. क इमं दर्शभिर्मिमेन्द्रं क्रीणाति धेनुमिः |झ 4 |24 |10

6. महेचन त्वाविः परा शुल्काय देयम् न सहस्राय नाप्रताय वाजिनों न शताय शतामथ। श प्रा.।

7. ऋ. वे. 6 |28 |5

8. गावो में अग्रतः सन्तु गावों मे सन्तु पृष्ठतः गावों में ह्रदयःसन्ततु गवां मध्ये वसाम्यहम्।

9. यज्ञाद् भवति पर्जन्यः) पर्जन्यारन्न सम्भवः।

10. अन्न प्राण बल चान्न अन्न सर्वार्थ साधनम् दूवासुर मनुष्याश्च सर्वे चान्नोय जीविनः।। कृ. वा.

11. कर्णे च हस्ते च सुवर्ण विद्यते यदिउपवासस्यथपि स्यादन्तामेवन देहिन्नम्।।

12. द्र. कृषिपाराशर।

13. गोमूत्रं गोमयं क्षीरं दधि सर्विः मुशोदकाम। निर्दिष्ट प०चगव्यं तु पवित्रं मुनि पुवैः।

14. आयुः कामों घृतनाग्गो जुहुयल्लिक्ष संख्यय।

15. घृत मानय पौराणं बीजिका लवणं दधि ब्रणभंत्र करिष्याकि स्वयमेव पिनाकिनः।। वामन पुराण।

16. बुद्धग्नि शुक्रौजो मेदः स्मृति कफ कारित्वम्। वातपित्त विषोन्माद रोग शोथालक्ष ज्वर नाशित्वम्।।

17. बालवृद्धिनां प्रशस्तत्वम्, बल कारित्वम्

18. दुग्धोत्थं नवनीतं तु चाक्षुष्यं रक्तपित्तनुत्।

19. रात्रौदधि न म०जीत।

20. समुद्धित घृतं तक्रं अर्धोद्धृत घृतं तक्ष। अनुद्घृत घृतं चान्यत् इयत् त्रिविध ां मतम्।

21. ससरं निर्जलं घोलं मथितं त्वसरोदकम। तक्रं पदिजल प्रोक्तं उदस्वित् अर्धवारिकम्।। छच्छिका सरहीन स्यात् अच्छा प्रचुरविम्।।

22. सतित्लं गुडसम्मिश्रं अ०जलयर्ध मितं प्यः। मार्कण्डेयाद्वरलब्ध्वा पिवाम्यायुर्विबृद्धये।।

23. स्वजन्म दिवसे यस्यु पायसेर्स धुरान्वितैः। जुहोति तस्य वर्धन्ते कमलारोग्य कीर्तयः।।

24. अमृतं शिशिरे बलिः अमृतं क्षीर भोजनम्।

25. आसपूरात्रं प्रसवात् क्षीर पेयूष मुह्यते। परतो मोरटं विद्याक अस्रन्न कफात्मकम्।। ৳

अध्याय 24

आयुर्वेदिक एवं स्वास्थ्य के सन्दर्भ में दुग्ध एवं दुग्ध-उत्पादों का महत्व

जी०सी० कात्यायन

सम्पूर्णानन्द संस्कृत विश्वविद्यालय

वाराणसी

आयुर्वेद क्या है? यह दो शव्दों आयु तथा वेद से वना है। यहाँ हम संक्षिप्त रूप से आयु की व्याख्या करेंगे। आयु शव्द के पर्याय निम्नलिखित हैं—

शरीरेन्द्रिय सत्वात्मसंयोगो धारि जीवितम्।[1]

नित्यगुश्चानुवन्धश्च पर्यायेरायुरुच्यते।।

वेद चार है ऋग, यजुर, साम तथा अथर्व, चौथे वेद अथर्ववेद का उपवेद आयुर्वेद है। भारत में आयुर्वेद अर्थात् जीवन विज्ञान के नाम से जाना गया ढांचा, जिसका विकास एवं वृद्धि भारतीय सभ्यता एवं संस्कृति की वृद्धि एवं मूल्यांकन के साथ–साथ हुई। चारों, वेद जिन्हें अभिलेखित भारतीय संस्कृति का खजाना माना जाता है, में इस जीवन विद्या का लिखित एवं व्यावहारिक ज्ञान दोनों रूपों में उपलब्ध हैं। इनमें जीवन्त तथा गैर जीवन्त सामग्री के गठन, मानव के शारीरिक, जैव रसायनिक, जैविक मनोवैज्ञानिक एवं आध

यात्मिक घटकों तथा मानव शरीर के अन्दर तथा शरीर के वाहर दोनों जगह काम करने वाली सजीव प्रेरणा शक्तियों के वारे में सिद्धान्तों की प्रस्तुति है। अन्य प्राचीन कार्यों में शरीर रचना विज्ञान, शरीर क्रिया विज्ञान, रोग हेतु विज्ञान, रोग विज्ञान, विकृति विज्ञान, उपचार ज्ञान, एवं पर्यावरणीय घटकों जैसे वर्तमान चिकित्सीय विषयों का वर्णन है।

चिकित्सा सम्वन्धी ज्ञान की प्राप्ति का कार्य युग–युगान्तर से होता आया है। संरक्षण की यह महान शक्ति, व्यापकीकरण हजारों वर्षो से किए गये अन्वेषकों के निरन्तर श्रम से सम्वद्ध विश्लेषण का परिणाम है। इस ज्ञान ने भारतीय संस्कृति के विकास में इतनी महत्वपूर्ण भूमिका का निर्वहन किया है कि इसका वेदों में, जो देवताओं से उत्पन्न हुए माने जाते हैं, में एकीकृत रूप में प्रमाण दिया गया है। आयुर्वेद में वर्णित अधिकांश पौराणिक तथा चिकित्सा सम्वन्धी धार्मिक सिद्धान्तों की उत्पत्ति आज भी प्राचीन काल का गुप्त विषय वना हुआ है। आयुर्वेद रूपी जीवन विज्ञान, जिसे आधुनिक समाज लगभग 1500 ईशा पूर्व का मानता है, में मौलिक तथा व्यावहारिक सिद्धान्त सूत्र रूप में संगठित एंव प्रतिपादित किया गया है। आयुर्वेद का मुख्य वेद अथर्ववेद में रोगों के उपचार के लिए 114 स्तोत्र या सूत्र हैं। इस भाव में आयुर्वेद को सकारात्मक स्वास्थ्य और मानव–वीमारी के उपचार के एक प्राचीनतम संगठित चिकित्सा पद्धति माना जाता है।

हजारों वर्षों से विधिवत् सतक्र अनुवीक्षण का प्रयोग करके तथा विस्तृत अनुभव का प्रमाण देने के लिए आयुर्वेद दो प्रमुख स्कूलों तथा आठ विशेषताओं वाली एक व्यापक स्वास्थ्य परिचर्चा प्रणाली के रूप में विकसित है। इनमें क्रमशरु आत्रेय सम्प्रदाय "वीववस व" चिलेपमपंदे तथा धनवन्तरि सम्प्रदाय "वीववस व" नतहमवदे के रूप में जाना जाता है जिसे शल्य चिकित्सा भी कहते हैं। चिकित्सा सिद्धान्त के अत्यधिक महत्वपूर्ण तथा महाकाय प्राचीन संकलनको चरक संहिता कहते हैं। इसमें थेराप्युटिक और आन्तरिक औषधि का प्रवर्धन करने सम्वन्धी विषय निहित है। इसप्राचीन ग्रन्थ में लगभग 600 औषध पादप, पशुज (जैवीय) तथा खनिजों की उत्पत्ति का वर्णन है।

आयुवेद के आठ अंगरु–

1. आन्तरिक औषधि– कायचिकित्सा
2. वाल रोग विज्ञान – कौमारमृत्यम्
3. मनश्चिकित्सा–ग्रह विज्ञानम्
4. नेत्र चिकित्सा–शालाक्य तंत्रम्
5. शल्य चिकित्सा–शल्यतंत्रम्
6. विष विज्ञान–विषतंत्र
7. जराचिकित्सा–रसायनतंत्रम्
8. वाजीकरण तंत्रम्

आयुर्वेद के चिकित्सा सिद्धान्त

1. पंचमहाभूत–मानव तथा अन्य जीवन की संरचना इन्हीं तत्वों से है।
2. त्रिदोष – प०चमहाभूत से ही वात, पित्त एवं कफ से तीन त्रिदोष उत्पन्न होते हैं।
3. सप्तधातु– रस,रक्त, मांस, मेदा, अस्थि, मज्जा तथा शुक्र इन्हीं सप्त धातुओं से शरीर का निर्माण हाता है।
4. मल विज्ञान– स्वेद, शकृत् तथा मूत्र ये तीन अपवर्ज्य शरीर में होते हैं।

आयुर्वेद के संक्षिप्त वर्णन के वाद अव मुख्य विषय दुग्ध उत्पादों पर आपका ध्यान आकर्षित करते हैं–

गाय शव्द की व्युत्पत्ति–गच्छति लोक कल्याणाय या सा गौरु। गमे 'र्डो', गमधातु से डो प्रत्यय करने पर गो शव्द वनता है।

ऋग्वेद में गाय के नौ नाम है।–अघ्न्या, उस्रा, उस्रिया, अही, मही, अदिति,इला, जगती तथा शकरी। भाव प्रकाश निघण्टु नामके आयुर्वेद के प्राचीन ग्रन्थ में दूध शव्द के पर्याय तथा गुण निम्नवत् वताये गये हैं–

दुग्धं क्षीर पयरुस्तन्यं वाल जीवनमित्यपि।

दुग्धं सुमधुरं स्निग्धं वातपित्तहरं सरम्।।[2]

अर्थात दुग्ध, क्षीर, पय (पयस्) , स्तन्य एवं वालजीवन। इसका गुण– मधुर रस युक्त, स्निग्ध, वात तथा पित्त को दूर करने वाला, सारक, तत्काल शुक्र को उत्पन्न करने वाला, शीतल, सम्पूर्ण प्राणियों के लिए सात्म्य (अनुकूल), जीवनी शक्ति को देने वाला, वृहण (रस रक्त आदि को वढ़ाने वाला), वल कारक, मेधा शक्ति के लिए हितकर, अत्यन्त वाजीकरण, अवस्था को स्थिर रखने वाला, आयु को वढ़ाने वाला, सन्धान कारक तथा रसायन है।

नौ प्रकार के दूधरु

आयुर्वेद में नौ प्रकार दूधों का वर्णन है–

1. गाय का दूध – (अ)सद्यरु प्रसूता गाय का दूध(व) वष्कयिण्या (वाखरी, वकेन) गाय का दूध। अन्य वर्ण की अपेक्षा के दूध के लिए काली गाय का दूध अधिक गुणकारी होता है।
2. भैंस का दूध
3. वकरी का दूध
4. जांगल देश की हरिणियों का दूध
5. भेंड़ी का दूध
6. घोड़ी का दूध
7. उँटनी का दूध
8. हथिनी का दूध
9. नारी दूध

इन सभी दूध में गाय का दूध स्वास्थ्य की दृष्टि से आयुर्वेद में सर्वोत्तम कहा गया है–गाय के दूध के गुण–गव्यं दुग्धं विशेषेण मधुरं रस पाकयोरु। दोष धातु मल स्रोतरु किंचिद् क्लेदकरं गुरु।।शीतलं स्तन्यकृस्निग्धं वात पित्तास्र नाशनम्। जरा समस्त रोगाणां शान्ति कृत सेविनां सदा।।[3]

अर्थात् गाय का दूध विशेष कर स्वाद तथा विपाक में मधुर रस युक्त, शीतल, दुग्ध वर्धक, स्निंग्ध, वात पित्त तथा रक्त विकार को नष्ट करनेवाला, दोष, धातु, मल तथा नाड़ियों में किंचित् क्लेद (आर्द्रता) पैदा करने वाला, गुरु एवंम् निरन्तर सेवन करने वालों की वृद्धावस्था तथा समस्त रोगों को शमन करने वाला होता है।

देश विशेष से गाय के दूध का गुण

जांगलानूप शैलेषु चरन्तीनाम् यथोत्तरं।

पयोगुरुतरं स्नेहों यथाऽऽहारं प्रवर्तते।।[4]

जांगल देश, आनूप देश, तथा पहाड़ों पर चरने वाली गायों का दूध उत्तरोत्तर एक दूसरे की अपेक्षा अधिक गुरु होता है। इसी तरह आहार विशेष से भी गाय के दूध का गुण परिवर्तित हो जाता है।

स्वल्पान्न भक्षणा ज्जातं क्षीरं गुरु कफ प्रदम्।

तत्तु वल्यं परं वृष्यं स्वस्थानाम गुण दायकम्।।

पलाल तृण कार्पास वीजजं रोगिणे हितम्।।[5]

जो गाय चारे के साथ थोड़ा अन्न भी खाती है, ऐसी गाय का दूध गुरु, कफ कारक, वल दायक तथा अत्यन्त वृष्य (वीर्य वर्द्धक) होता है। अतएव वह स्वस्थ लोगों के लिए गुण दायक होता है तथा जो गाय पैरा (भूसा), घास तथा कपास का वीज खाकर दूध देती है–ऐसी गाय का दूध रोगियो के लिए हितकर होता है।

सुश्रुत संहिता के अनुसार वाल्यावस्था में प्रथम दो–तीन वर्ष तक दूध जीवन का खास आधार है, संसार में दूध के समान अन्य कोई खास द्रव्य नहीं है। जन्म से मृत्यु पर्यन्त किसी भी दशा में दुग्ध निषिद्ध नहीं है। शरीर वर्द्धन

के लिए जितने उपादान आवश्यक हैं वे सव दुग्ध में मौजूद रहते हैं। इसीलिए दुग्ध को पूर्ण आहार माना गया है। दूध में प्रोटीन, वसा या चर्वी, दुग्ध शक्ररा (व्जेंम), चूना, सोडियम पोटेशियम, मैग्नीशियम, लौह, गन्धक, फास्फोरस, आयोडीन तथा सभी विटामिन (विशेषतया ए0 डी0 ई0) होते हैं।

दुग्ध शव्द की व्युत्पत्ति

दुह्यते निरु श्रेयसाय यत् तद् दुग्धम्।

अर्थात् जो निरुश्रेयस (यज्ञादि) के लिए दुहा जाय वह दुग्ध है।

गाय के लिए महाकवि भरवि ने कहा है–

जगत प्रसूतिर्जगदेकपावनी"।

महर्षि सुश्रुत के अनुसार–

अल्पाभिष्यान्दि गोक्षीरं स्निग्धं गुरु रसायनम्।

रक्त पित्त हरं शीतं मधुरं रस पाकयोरु।।

जीवनीयं तथा वातपित्तध्न परम् स्मृतम्।।[6]

अर्थात् अल्प अभिष्यन्दि कारक, स्निग्ध, गुरु, रसायन, रक्त पित्त हारक, शीत, रस और पाक में मधुर, जीवनीय तथा वात पित्त नाशक है।

प्रातरु एवम् शायम् कालीन दुग्ध में अन्तररु

महर्षि सुश्रुत ने प्रातरु तथा शायंकालीन दुग्ध में निम्न प्रकार से अन्तर उपदेशित किया है–

प्रायरु प्राभातिकं क्षीर गुरु विष्टम्भि शीतलम्।

रायारु सोम गुणत्वा०च व्यायामाभावतस्तथा।।

दिवाकराभि तप्तानां व्यायामानिल सेवनात्।

वातानुलोमि श्रान्तिघ्नं चक्षुष्यं चापराह्निकम्।।[7]

वहुधा प्रातरु काल में निकाला हुआ दूध गुरु, विष्टम्भ कारक तथा शीतल होता है क्योंकि रात्रि के सोम (चन्द्रमा) गुण प्रधान होने से तथा जानवरों के चलने फिरने रूपी व्यायाम के न होने से प्रभात कालीन दुग्ध गुर्वादि गुणों वाला होता है परन्तु दिन में पशुओं के सूर्य की धूप में चरने से, व्यायाम और वात का सेवन होने से सन्ध्या समय का दुग्ध वात का अनुलोमक (नाशक), थकावट को दूर करने वाला तथा नेत्रों के लिए हितकर होता है।

कच्चा दुग्ध प्रायरु अभिष्यन्द (क्लेद) कारक तथा पचने में भारी होता है किन्तु उसी को उवाल लेने (गरम करने) से अति हल्का तथा अभिष्यन्द नहीं करने वाला हो जाता है। स्त्रियों के दुग्ध को छोड़कर सभी प्रकार के दुग्धों को गरम करके पीना चाहिए। क्योंकि स्त्री का दूध कच्चा ही पीने से हित करता है।

पयोभिष्यन्दि गुर्वाग्न प्रायशरु परिकीर्तितम्।

तदेवोक्तं लघुतरमनभिष्यन्दि वै श्रृतम।।

वर्जयित्वा स्त्रियारु स्तन्यमाममेव हि तद्धितम्।।[8]

चरक संहिताकार महर्षि चरक ने सर्व प्रथम गाय के दुग्ध में दस गुण होना वताया है, इसका वर्णन निम्न तरह से किया है—गाय के दुग्ध में दस गुण—

स्वादुशीतं मृदु स्ग्धिं वहलं श्लक्ष्ण पिच्छिलम्

गुरु मन्दं प्रसन्नं च गव्यं दशगुण पयरु।।

तदेवं गुणमेवौजरु सामान्यादभिवर्धयेत्।

प्रवरं जीवनीयानामक्षीरमुक्तं रसायनम्।।[9]

गाय के दुग्ध के गुण

रस में स्वादु, वीर्य में शीत, गुण में मृदु, स्निग्ध, वहल (गाढ़ा) श्लक्ष्ण,पिच्छिल, गुरु, मन्द तथा प्रसन्न। इन दश गुणों से युक्त दुग्ध इन्हीं दस गुणों वाले ओज का सात्म्य रखने के कारण उसकी वृद्धि करता है। यह दूध जीवनी शक्ति

प्रदान करने वाले द्रव्यों में सवसे श्रेष्ठ और रसायन है। दूध के आयुर्वेदशास्त्रीय विवेचन के वाद दूध के अन्यान्य उत्पादों के शास्त्रीय पक्ष का वर्णन–

दूध का प्रथम उत्पाद दही (दधि)

महर्षि चरक के अनुसार दही के गुण–

रोचनं दीपनं वृष्यं स्नेहनं वल वर्द्धनम।

पाकेऽम्लमुष्णं वातघ्नं मंगल्यं वृंहणं दधि।।[10]

यह दही भोजन में रुचि उत्पन्न करेवाला, अग्नि को दीप्त करता है, शुक्रवर्द्धक, शरीर में स्निग्धता लाता है, वल वर्द्धक है, विपाक में खट्टा है, वीर्य में उष्ण, वातनाशक, मंगलकारी, मास को वढ़ाने वाला होता है।

पीनसे चातिसारे च शीतके विषमज्वरे।

अरुचौ मूत्रकृच्छे च कार्श्ये च दधि शस्यते।।

शरद ग्रीष्म वसन्तेषु प्रायशों दधि गर्हितम्।

रक्त पित्त कफोत्थेषु विकारेष्वहितं च तत्।।

त्रिदोषं मन्दकं,जातं वातघ्नं दधि, शुक्रलरु।

सररु श्लेष्मानिलघ्नस्तु मण्डरु स्रोतोविशोधनम्।।[11]

(चरक संहिता सूत्र अध्याय 27)

दही का चिकित्सकीय उपयोग

पीनस, अतिसार, शीत लगकर आने वाला ज्वर, विषम ज्वर, अरुचि, मूत्रकृच्छ तथा शरीर के कृश होने पर दही का सेवन करने से लाभ होता है।

प्रायरु शरद, ग्रीष्म तथा वसन्तु ऋतु में दही का सेवन करना अच्छा नहीं होता है।

रक्त पित्त एवं कफजन्य विकारों में दही खाना हानिकारक है।

मन्दक दधि (जो उत्तम रूप से जमा हुआ न हो) वात, पित्त तथा कफ को कुपित करता है।

जात दधि (उत्तम रूप से जमा हुआ) वात नाशक होता है।

दही की मलाई शुक्रवर्धक है।

दही का जल, कफ तथा वातनाशक एंव स्रोतों को शुद्ध करता है।

महर्षि सुश्रुत ने अपनी संहिता में दही के वारे में निम्न प्रकार से वर्णन किया है– सुश्रुत संहिता के सूत्र स्थान के अध्याय 45 में,

दधि तु मधुरम् अम्लं अतिअम्लंचेति, तत्कषायानुरसं स्निग्धमुष्णं पीनस
विषम ज्वरातिसार मूत्रकृच्छरु कार्श्यापहं वृष्यं प्राणकरं मंगल्यं च।[12]

अर्थात दही वहुधा मधुर अम्ल और अधिक अम्ल इस तरह तीन प्रकार का होता है, दही अनुरस में कषाय, स्निग्ध और उष्ण वीर्य होता है तथा पीनस, विषम ज्वर अतीसार, अरोचक, मूत्रकृच्छ एवं शरीर की कृशता को नष्ट करता है। वृष्य गुण युक्त, प्राण शक्ति को वढ़ाने वाला तथा पूजन आदि कार्यों में मांगलिक माना गया है।

वैज्ञानकि दृष्टि से

1. दही के क्षीराम्ल जीवाणु विसूचिका ;वैवससमतंद्ध के जीवाणुओं को नष्ट करते हैं।

2. दही आन्त्र में मल की सड़न को रोकता है। मेचनीकाफ नामक वैज्ञानिक ने यह सिद्ध किया है कि प्रतिदिन दही सेवन करने से आन्त्रस्थ जीवाणु तथ उनका विष नष्ट होकर वृद्धावस्था देर से आती है।

3. पाचन संस्थान के अनेक विकार जैसे मंदाग्नि एंव संग्रहणी में दही यातक्र अत्यन्त ळप्प्ज लाभकारी सिद्ध हुआ है।

महाभिष्यन्दि मधुर कफ मेदो विवर्द्धनम्।

कफपित्तकृदम्लं स्यादत्यम्लं रक्त दूषणम्।

विदाहि सृष्ट विण्मूत्रं मन्दजातं त्रिदोषकृत्।।[13]

अर्थात मीठा दही अत्यन्त अभिष्यन्द (क्लेद) कारक और कफ तथा मेदोधातु का वर्द्धक होता है, खट्टा दही कफ और पित्त कारक होता है एवं अधिक खट्टा दही रक्त को दूषित करता है। मन्दजात (ठीक–ठीक न जाम हुआ दही) शरीर में विदाह करने वाला, विष्ठा और मूत्र को पैदा करने वाला या उनका निरुस्सारक एवं त्रिदोष कारक होता है।

गाय का दही

स्निग्धं विपाके मधुरं दीपनं वल वर्द्धनम्।

वातापहं पवित्रं दधि गव्यं रुचि प्रदम्।।[14]

अर्थात गाय का दही स्निग्ध, विपाक में मधुर, अग्नि दीपक, वलवर्द्धक, वात कानाशक, पवित्र और रुचि कारक होता है। गाय, वकरी, भैंस, ऊंटनी, भेड़, घोड़ी, स्त्री तथा हस्तिनी इन 8 प्रकार की दही में महर्षि सुश्रुत ने गाय की दही को स्वास्थ्य की दृष्टि से सर्वोत्तम वताया है।

वातघ्नं कफ कृत स्निग्धं वृंहणं नाति पित्तकृत्।

कुयदि् भक्ताभिलाषंच दधियत् सुपरिस्रुतम्।।[15]

कपड़ें में वाँधकर अच्छी प्रकार परिस्रुत किया हुआ दही वातनाशक, कफ कारक, स्निग्ध तथा शरीर की पुष्टि (वृंहण) करने वाला होता है तथा यह दही पित्त को अधिक नहीं वढ़ाता है एवं भोजन के प्रति रुचि वढ़ाता है।

श्रृतात क्षीरात्तु यज्जातं गुणवत् दधि तत् स्मृतम्।

वात पित्त हरं रुच्यं धात्वग्नि वल वर्द्धनम्।।[16]

अच्छी तरह उवाले हुए दूध से जो दही वनता है वह अधिक गुणकारी,वात तथा पित्त का नाशक, वचि कारक एवं शरीर की धातुओं, अग्नि तथा वल को वढ़ाता है।

दध्नरु सरो गुरुर्वृष्यो विज्ञेयोऽनिल नाशनरु।

वह्नेर्विधमनंचापि कफ शुक्र विवर्द्धनरु।।[17]

दही के ऊपर का भाग (सर) पचने में भारी, वृष्य, वात का नाशक, पाचकाग्नि का नाशक तथा कफ और शुक्र को वढ़ाता है।

सर अर्थात् स्नेह वहुल ऊपरितनांशरु।''

दधित्वसारं रूक्षंच ग्राहि विष्टंभि वातलम्।

दीपनीयं लघुतरं सकषायं रुचि प्रदम्।।[18]

जिसमें से मक्खन या क्रीम निकाल लिया गया हो ऐसे दूध का जमाया हुआ दही रूक्ष, ग्राही, विष्टम्भि, वातजनक, अग्नि दीपक, पचने में अत्यन्त हल्का, कषाय रस युक्त तथा रुचिप्रद होता है।

महर्षि सुश्रुत के अनुसार–

शरद् ग्रीष्म वसन्तेषु प्रायशो दधि गर्हितम्।

हेमन्तें शिशिरेचैव वर्षासु दधि शस्यते।।[19]

शरद ग्रीष्म तथा वसन्त ऋतुओं में प्रायरु दही अधिक खाना ठीक नहीं है। हेमन्त, शिशिर और वर्षा ऋतु में दही का सेवन श्रेष्ठ होता है।

तृष्णा क्लमहरमस्तु लघु स्रोतो विशोधनम्।

अम्लं कषायं मधुरं वृष्यं कफ वातनुत्।।

प्रह्णादनं प्रीणनं च भिनत्याशु मलं च तत्।

जलमावहते क्षिप्रं भक्तच्छन्दं करोति च।।[20]

दही का पानी (मस्तु) तृष्णा, क्लम को नष्ठ करता है। पचने में लघु, स्रोतसों का विशोधक तथा अम्ल, कषाय और मधुर होता है। यह वृष्य नहीं है, कफ एवं वात को नष्ट करता है। आत्मा, मन और इन्द्रियों को आह्लादित करता है तथा तृप्त करता है मल का शीघ्र भेदन करता है। शीघ्र शरीर के वल को वढ़ाता है और भोजन में रुचि लाता है।

भाव प्रकाश निघटुकार भावमिश्र ने दही के वारे में निम्न प्रकार से वर्णन किया है–

दध्युष्ण दीपनं स्निग्धं कषायानुरसं गुरु।

पाकेम्लं ग्राहि पित्तास शोथ मेदरु कफ प्रदम।।

मूत्र कृच्छ्रे प्रतिश्याये शीतगे विषमज्वरं।

अतीसारेऽअरुचौ कार्श्ये शस्यते वल शुक्रकृत्।।[21]

दही उष्ण, अग्निदीपक, स्निग्ध, किंचित कषाय रस युक्त, गुरु विपाक में अम्ल रस युक्त, ग्राही एवं पित्त रक्त विकार, शोथ, मेद तथा कफ को उत्पन्न करने वाला है। चिकित्सा में दही का उपयोग मूत्रकृच्छ्र, जुकाम, शीत, विषमज्वर, अतीसार, अरुचि तथा कृशता में अच्छा कार्य करता है। यह वल तथा शुक्र को वढ़ाने वाला होता है।

दही के प्रकार

1. मन्द
2. स्वादु
3. स्वाद्वम्ल
4. अम्ल
5. अत्यम्ल

गाय के दही का गुण

गव्यं दधि विशेषण स्वाद्वम्लं च रुचि प्रदम्

पवित्रं दीपनं हृद्यं पुष्टि कृत्पवनापहम्।।

उक्त दध्नामशेषाणाम् मध्ये गव्यं गुणाधिकम्।[22]

गाय का दही विशेष रूप से मधुर तथा अम्ल रस युक्त, रुचि पैदा करने वाला, पवित्र, अग्नि दीपक, हृदय के लिए हितकर, पुष्टिकारक तथा वात नाशक होता है तथा सम्पूर्ण दहियों के वीच में गाय का दही सर्वोत्तम गुणकारी है।

रात में दही का सेवन निषिद्ध है–

न नक्तं दधि भुंजीत न चाप्य घृत शक्ररम्

ना मुद्‌ग सूपं नाक्षौद्रं नोष्णं नामलकैर्विना।।[23]

रात में दही नहीं खाना चाहिए और विना घी तथा शक्कर के, विना मूंग की दाल के, विना मधु के, विना आंवला के दही नहीं खाना चाहिए।

तात्पर्य यह है कि यदि रात में दही खाना पड़े तो घी, शक्कर आदि से युक्त दही ही खायें।

ज्वरासृक्पित्त विसर्प कुष्ठ पाण्ड्वामय भ्रमान्।

प्राप्नुयात् कामलां चोग्रां विधिं हत्वा दधिप्रियरु।।[24]

उपरोक्त निषिद्ध समय में जो व्यक्ति दही का सेवन करता है उसे ज्वर, रक्त, पित्त, पिसर्प, कुष्ठ, पाण्डु तथा भ्रम रोग एवं प्रचण्ड रूप से कामला रोग उत्पन्न हो जाता है सर (साढ़ी) दही के ऊपर जो गाढ़ा तथा स्नेह (घी) से युक्त भाग है उसे सर (साढ़ी) कहते हैं।

मस्तु के लक्षण

दही के माण्ड (पानी) को मस्तु या दही का तोड़ कहते हैं।

तक्र

चरक संहिता के सूत्र स्थान के अन्नपान्न विधि अध्याय नामक 26 वें अध्याय में तक्र (मठ्‌ठा) के वारे में विचार इस प्रकार है–

शोफार्शो ग्रहणी दोष मूत्र ग्रहोदरा रुचौ।

स्नेह व्यापदि पाण्डुत्वे तक्रं दद्याद्‌गरेषु च।।[25]

यह मठ्‌ठा शोथ, अर्श, ग्रहणी, मूत्रकृच्छ्र, उदर तथा अरुचि रोग में, स्नेह पीने के वाद होने वाले उपद्रव में, पाण्डु रोग में तथा विष विकार में उत्तम दवा का कार्य करता है।

सुश्रुत संहिता में तक्र आदि का वर्णन इस प्रकार है–

मन्थनादपृथग्भूत स्नेहमद्धोर्दकं च यत्।

नातिसान्द्र द्रवं तक्रं स्वाद्धम्लं तुवरं रसे

यत्तु सस्नेहमजलं मथितं घोलमुच्यते।[26]

मन्थन करके जिसमें से (स्नेह) मक्खन निकाल लिया गया हो और दही से आधा पानी मिला हुआ हो, जो न अधिक गाढ़ा और न अधिक पतली हो तथा जो रस में मधुर, अम्ल और कषाय हो उसे तक्र कहते हैं किन्तु जिस दही में से मक्खन न निकाला गया हो तथा पानी भी नहीं डाला गया हो, केवल मथनी से घोल (विलयन) वना दिया गया हो उसे घोल कहते हैं।

तक्र के तीन प्रकार

रूक्षमर्द्धोद्घृतं स्नेहं यतश्चानुधृतं घृतम्।

तक्रं दोषाग्नि वलविस्यिविधं तत्प्रयोजयेत्।।[27]

तक्र सेवन में निषेध

नैव तकं क्षते दद्यान्नोष्ण काले न दुर्वले।

न मूर्च्छाभ्रमदाहेषु न रोगे रक्तपैत्तिके।। 28

क्षत्, गरमी का समय, दुर्वल मनुष्य, मूर्च्छा, भ्रम, दाह तथा रक्तपित्त के रोगी इन्हें तक्र का प्रयोग नहीं करना चाहिए।

तक्र सेवन का प्रशस्त समय

शीतकाले अग्निमांद्य च कफोत्थेष्वामयेषु च।

मार्गावरोधे दुष्टे च वायौ तक्रं प्रशस्यते।।[29]

शीत काल, अग्निमाद्य, कफ जन्य रोग, स्रोतसों के मार्गावरोध तथा वायु की दुष्टि में तक्र प्रशस्त है।

वात के प्रकोप में सैन्धव लवण युक्त अम्लतक्र, पित्त के प्रकोप में शक्ररायुक्त स्वादु तक्र तथा कफ के प्रकोप में सोठ, मरिच, पिपपली तथा यवक्षार इनका चूर्ण मिलाकर तक्र पीना चाहिए।

'तक्र शक्रस्य दुर्लभम' इन्द्र तक को तक्र नहीं उपलब्ध होता है, अर्थात पृथ्वी के मानव ही इसे प्राप्त करते हैं।

न तक्र सेवी व्यथते कदा्चिन्न तक्र दग्धारु प्रभवन्ति रोगारु।

यथा सुराणांममृतं हिताय तथा नराणां भुवि तक्र माहुः।।[30]

लैक्टिक एसिड वनाने वाले जीवाणुओं की क्रिया दुग्ध पर होने से दही बनता है, ये जीवाणु दुग्ध में अभिषंग ;थ्मतउमदजंजपवदद्ध उत्पन्न करके दुग्ध शक्ररा को लैक्टिकएसिड में परिवर्तित कर देते हैं। इस अम्ल से दूध की चरवी (मेद या स्नेहांश) और प्रोटीन जम कर दही वना जाता है।

भाव मिश्र ने भाव प्रकाश निघण्डु में तक्र वर्ग अध्याय में तक्र के गुण निम्न प्रकार से वर्णित किये हैं–

घोलं तु मथितं तक्र मुदश्विच्छच्छिकाऽपि च।

ससरं निर्जलं घोलं मथितं त्व सरोदकम्।।

तक्रं पादजलं प्रोक्तमुदश्वित्वर्द्धवारिकम्।

छच्छिका सारहीना स्यात् तु स्वच्छा प्रचुर वारिका।।

घोलं तु शक्ररायुक्तं गुणैर्ज्ञेयं रसालवत्।।[31]

तक्र (मठ्ठा) के भिन्न–भिन्न लक्षणों के अनुसार भिन्न–भिन्न संस्कृत नाम है–घोल,मथित, तक्र, उदश्वित तथा छच्छिका।

घोल के वारे में पूर्व में वताया जा चुका है। मथित– यदि दही की मलाई अलग कर विना जल मिलाये ही मथ दिया जाय तो उसे मथित कहते हैं।

तक्र

जिस दही में चतुर्थांश जल मिला कर मथा जाय उसे तक्र कहते हैं।

उदश्वित

जिस दही में आधा जल मिला कर मथा जाय उसे उद्श्वित् कहते हैं।

छच्छिका

जिस दही में से प्रथम मथ कर मक्खन निकाल लिया गया हो, पुनरु उसी में अधिक मात्रा में स्वच्छ जल डालकर फिर मथा जाय उसे छच्छिका कहते हैं।

घोलं तु शक्ररायुक्तं गुणैर्ज्ञेयं रसालवत्।

वात पित्त हरं ह्लादि मथितं कफ पित्तनुत्।।[32]

घोल में शक्कर मिलाकर सेवन करने से यह शिखरण के समान वात पित्त नाशक, आह्ल्लाद जनक होता है।

तक्रं–ग्राहि कषायाम्लं स्वादुपाक रसं लघु।

वीर्योष्णं दीपन वृष्यं प्रीणनं वातनाशनम्।।

ग्रहण्यादिमतां पथ्यं भवेत् संग्राहि लाघवात्।

किच स्वादु विपाकीत्वान्न च पित्त प्रकोपनम्।।

अम्लोष्णं दीपनं वृष्यं प्रीणनं वात नाशनम्।

कषायोष्ण विकाशित्वात् द्रौक्ष्यांष्चापि कफापहम्।।[33]

अर्थात तक्र कषाय मधुर रस युक्त विपाक मे मधुर ग्रहणी रोग में हितकर है। लघुहोने के कारण मलका संग्राहक है। तक्र का सेवन करने वाला कभी वीमार नहीं पड़ता। तक्र के प्रभाव से नष्ट हुए रोग पुनरु कभी उत्पन्न नहीं हो सकते।

उदश्वित् कफकृद् बलमामघ्नं परमं मतम्

छच्छिका शीतला लघ्की पित्त श्रम तृषा हरी।।

वातानुत् कफकृत् सातु दीपनी लवणान्विता।[34]

यह कफ कारक, वल वर्द्धक तथा अत्यन्त आम नाशक होता है। शीतल, लघु एवं पित्त, श्रम तथा तृषा को दूर करने वाला वात नाशक, कफ कारक होता है, यदि इसमें सेंधा नमक मिला हो तो यह अग्नि दीपक होता है।

दोष विशेष में तथा रोग विशेष में विशेष–विशेष तक्रों का प्रयोग

वातेऽम्लं शस्यते तक्रं शुषठी सैन्धव संयुतम्।

पित्ते स्वादु सितायुक्तं व्योषक्षार युतं कफे।।[35]

वात दोष की अधिकतां में अम्ल रस युक्त एवम् सोठ तथा सेंधा नमक मिला हुआ तक्र अधिक हितकर है।

पित्त की अधिकता में मधुर रस युक्त तथा चीनी मिश्रित तक्र उत्तम हितकारी होता है। कफ की अधिकता में सोठ, मरिच, तथा पीपर मिश्रित तक्र उत्तम हितकारी होता है।

हिंगु जीर युतं घोलं सैन्धवेन च संयुतम्।

भवेदतीव वातघ्नमर्शोऽतीसार हृत्परम्।।

रूचिदें पुष्टिदं वल्य वस्ति शूल विनाशनम्।

मूत्रकृच्छ्रे तु सगुडं पाण्डुरोगे सचित्रकम्।[36]

हींग, जीरा (ये दोनो भुने हुए हों) तथा सेंधा नमक से युक्त घोल–अत्यन्त वातनाशक, अर्श (पाइल्स) अतीसार (डिसेन्ट्री) रोग को ठीक करने वाला, रूचि कारक, पुष्टिकारक, वलकारी तथा वस्ति शूल नाशक होता है। गुड़ मिला घोल– मूत्रकृच्छ्र (Urinary retention) में चित्रक (चीता) मिश्रित घोल पाण्डुरोग (Aoitis) में देना हितकर होता है।

तक्रमामं कफं कोष्ठे हन्ति कण्ठे करोति च।

पीनसं श्वास कासादौ पक्वमेव प्रयुज्यते।।[37]

कच्चा (विना पकाया हुआ) तक्र, कोष्ट (G.I.T.) स्थित कफ को नष्ट करता है, किन्तु कण्ठ में कफ करने वाला होता है। अतरु पकाए हुए तक्र का

ही – पीनस, श्वास तथा कास आदि में प्रयोग करना उचित है क्योंकि हितकर होता है।

रोग के अनुसार तक्र सेवन करें

शीत काल में तथा अग्नि की मन्दता (Indigetion), वात रोग, अरुचि (Anorexia), नाड़ियों के अवरोध में तक्र अमृत की तरह गुणकारी होता है। तक्र गर विष (संयोगज विष), वमन (Vomiting), प्रसेक, (कफ जन्य लार आदि गिरना), विषम ज्वर (Typhoid), पाण्डुरोग (Aoitis), मेद रोग (Fatiness), ग्रहणी, अर्श (Piles बवासीर), मूत्रग्रह (मूत्र का वन्द होना), भगन्दर (Fistulla), प्रमेह (Dibetes), गुल्म, अतीसार, शूल, प्लीहा, उदररोग, अरुचि, श्वित्र (श्वेत कुष्ठ), कोष्ठगत रोग, शोथ, तृषा तथा क्रिमि रोग को नष्ट करता है।

किन रोगो के रोगी को तक्र निषेध है–

नैव तक्रं क्षये दद्यात् नोष्ण काले न दुर्वले।

न मूर्च्छा भ्रम दाहेषु न रोगे रक्तपित्तजे।।[38]

क्षय रोग (थाइसिस) में, ग्रीष्म ऋतु में एवम दुर्वल व्यक्ति तथा मूर्च्छा, भ्रम, दाह तथा रक्त पित्त रोग युक्त व्यक्ति को नहीं देना चाहिए। तात्पर्य यह है कि ऐसे व्यक्ति को तक्र सेवन अहितकर है।

मक्खन:

महर्षि चरक ने अपने विश्व प्रसिद्ध ग्रन्थ 'चरक संहिता' के सूत्र स्थान के अन्नपान विषयक 27 वे अध्याय में मक्खन के विषय में कहा है कि–

संग्राहि दीपनं ह्रद्यं नवनीतं नवोद्घृतम्।

ग्रहण्यर्शो विकाराघ्नमर्दिता रुचिनाशनम्।।[39]

नवीन मक्खन मल को वाधने वाला, अग्नि को तीव्र करने वाला तथा ह्रदय के लिए हितकारी होता है। ग्रहणी एवं अर्श को दूर करता है। अर्दित थ्वपंस च्तंसलेपे अरुचि ।दवतमगपं को नष्ट करता है।

सुश्रुत संहिता Sohool of Sergeon के महान चिकित्सा वैज्ञानिक महर्षि सुश्रुत ने मक्खन (नवनीत) के विषय में कहा है कि

नवनीतं पुनः सद्यस्कं लघु सुकुमारं मधुरं कषायम्
ईषद् अम्लं शीतलं मेध्यं दीपनं ह्रद्यं .... ।[40]

(सुरुत संहिता सूत्रस्थान अध्याय 45से)

दही से ताजा निकाला हुआ मक्खन पचने में हल्का, शरीर को सुकुमार वनाने वाला रस में मधुर, कषाय, और कुछ खट्टा, शीत प्रकृतिक, मेधा वर्द्धक, अग्निदीपक,ह्रदय के लिए हितकारी, दस्त को वाँधने वाला, पित्त और वात का नाशक, वृष्य, दाह नहीं करने वाला, राजयक्ष्मा, कास, व्रण, व्यवाय–शोक–वार्द्धक्यादि जन्य शोष, अर्श तथा अर्दित थंबपंस च्तंसलेपे रोगों का विनाशक है। निकालकर रक्खा हुआ पुराना मक्खन पचने में भारी, कफ और मेदा को वढ़ानेवाला, बल वर्द्धक, वृंहण, शोष नाशक तथा अधिकतया वालकों के लिए प्रशस्त होता है।

भाव प्रकाश निघण्टु में मक्खन आदि पर विचारण

मक्खन के पर्याय

म्रक्षण, सरज, हैयंगवीन तथा नवीनतम्।

म्रक्षणं सरजं हैयंगवीनं नवनीतकम्।

नवनीतं हितं गव्यं वृष्यं वर्ण वलाग्निकृत्।।[41]

संग्राहि वात पित्तासृक्क्षयार्शोऽर्दित कास ह्रत्।

तद्धितं वालके वृद्धे विशेषादमृतं शिशो।।[42]

यह हितकर, वीर्यवर्द्धक, वर्ण (रंग) को उत्तम करने वाला, वल तथा अग्नि को वढ़ानेवाला, मल संग्राही है। वात, पित्त, रक्तगत विकार, क्षय, अर्श अर्दितवात तथा कास को दूर करने वाला होता है। यह वालक तथा वृद्ध के लिए हितकर एवम् विशेष करके शिशु (अति छोटे वच्चों) के लिए अमृत की तरह गुणकारी होता है।

दूध से निकला हुआ मक्खनः

दुग्धोत्थं नवनीतं तु चक्षुष्यं रक्त पित्तनुत्।

वृष्यं वल्यमति स्निग्धं मधुर ग्राहि शीतलम्।।[43]

यह नेत्रों के लिए हितकर, रक्त पित्त नाशक, वीर्य वर्द्धक, वल कारक, अत्यन्त स्निग्ध मधुर रस युक्त, ग्राही तथा शीतल होता है।

पुराना मक्खन–क्षार, कटु तथा अम्लरस युक्त होने से वमन, अर्श तथा कुष्ठ को पैदा करता है एवम् कफ जनक, गुरु तथा मेद को वढाने वाला होता है।

दधि कूर्चिका तथा तक्र कूर्चिका

दध्ना सह च यत्पक्वं क्षीरं सा दधि कूर्चिका।

तक्रेण पक्वं यत्क्षीरं सा भवेत् तक्र कूर्चिका।।[44]

तक्र कूर्चिका कब्ज करने वाली, वात कारक, रूक्ष और दुर्जर होती है। कूर्चिका भूत दही और तक्र से वनाया हुआ मण्ड तक्र से अधिक लघु (हल्का) होता है।

किलाट

गुरुः किलाटोऽनिलहा पुंस्त्व निद्राप्रदःस्मृतः।

मधुरौ वृंहणौ वृष्यौ तद्वत्पीयूष मोरटौ।।[45]

किलाट पचने में भारी, वातनाशक, पुंसत्व तथा निद्रा लाने वाला होता है। पीयूष (फेनुस)तथा मोरट स्वाद में मधुर, वृंहणकारक और वृष्य होते हैं

किलाटरु–कूर्चिकी भूत दुग्धस्य पिण्डः।

नष्ट दुग्धस्य पक्वस्य पिण्डःप्रोक्तःकिलाटकः।।

पीयूष–क्षीर सद्यःप्रसूतायाःपीयूषमिति संज्ञितम्।

मोरटः–सप्त रात्रात्परं क्षीरम्प्रसन्नस्तु मोरटः।

सन्तानिका—(दूध के ऊपर की मलाई)

सन्तानिका पुनर्वातघ्नी तर्पणी वल्या वृष्या स्निग्धा रुच्या मधुरा मधुर विपाका रक्त पित्त प्रसादनी गुर्वी च। 46

यह वातनाशिका, तृप्ति कारिणी, वल के लिए हितकर, वृष्य, स्निग्ध, रुचि वर्द्धक स्वाद और पाक में मधुर, रक्त पित्त का प्रसादन (संशमन) करने वाली तथा पचने में भारी होती है।

घृतः दूध के विशेष उत्पाद घी के बारे में आयुर्वेदीय (चरक, सुश्रुत, भवप्रकाश) अवधारणा

चरक—

स्मृति वुद्धयग्नि शुक्रौजरु कफ मेदो विवर्धनम्।
वात पित्त विषोन्माद शोषालक्ष्मी ज्वरापहम।।[47]
सर्व स्नेहोत्तम शीतं मधुर रस पोकयोः।
सहस्रवीर्यं विधिभिर्घृतं कर्म सहस्रकृत।।[48]

सामान्यतरु घृत स्मरणशक्ति, वुद्धि, अग्नि, वीर्य, ओज कफ और मेदा को वढ़ाने वाला होता है। वात, पित्त विषजन्य विकार, उन्मादरोग, राजयक्ष्मा, शरीर की अशोभा और ज्वर को नष्ट करता है। यह सभी स्नेहों में श्रेष्ठ होता है।

वीर्य में शीत, रस और विपाक में मधुर होता है, विभिन्न द्रव्यों से संस्कारित होने पर इसमें हजार गुनी शक्ति आ जाती है, अतरु उसी के अनुसार हजार प्रकार के रोगों को दूर करने का गुण इसमें आ जाता है।

पुराने घी का गुण

मदापस्पारमूर्च्छायशोषोन्मादगरज्वरान्।
योनि कर्ण शिरःशूलं घृतं जीर्णं व्यपोहति।।[49]

पुराना घृत मद, मिर्गी, मुर्च्छा, यक्ष्मा, पागलपन, कृत्रिम विष के प्रभाव, ज्वर योनिशूल, कान की वेदना, शिर की वेदना को शान्त करता है। वकरी, भेड़, भैस आदि का घृत इनके दुग्ध के गुण के समान ही गुणवाला होता है।

घृत एक वर्ष के वाद पुराना कहा जाता है।

चरक ने दस वर्ष के घृत को पुराना कहा है–

उग्र गन्धं पुराणंस्यात् दश वर्ष स्थितं घृतम्।

लाक्षारस समं शीतं प्रपुराण मतरु परम्।।[50]

सुश्रुत ने 111 वर्ष तक के पुराने घृत को कुम्भसर्पि, तथा इससे और पुराने घृत को महासर्पि माना है–

एकादशशतश्चैव वत्सरानुषितं घृतम।

रक्षोघ्नं कुम्भ सर्पिस्यात् परतस्तु महाघृतम्।।[51]

सुश्रुत–विपाके मधुरं शीतं वातपित्त विषापहम्।

चक्षुष्यम् वल्यञच गव्यं सर्पिगुणोत्तरम्।।[52]

यह विपाक में मधुर, शीत प्रकृतिक, वात, पित्त और विष का नाशक, नेत्रों के लिए हितकारी, वल वर्द्धक होता है। यह उदावर्त, उन्माद, अपस्मार, शूल, ज्वर, आनाह, और वात तथा पित्त को शांत करने वाला, पाचकाग्नि का दीपक है एवं स्मृति, मति, मेधा, कान्ति, स्वर लावण्य, सौकुमार्य, ओज, तेज और वल इन सवको वढ़ाता है। विष को नष्ट करता है, सूक्ष्म जीवाणुओं का नाशक है।

भाव प्रकाश–घृत के पर्याय

घृत, आज्य, हविस् तथा सर्पिस्।

घृतमाज्यं हविरु सर्पिरु कथ्यन्ते तद्गुणा अथ।

घृतं रसायनं स्वाद चक्षुष्यं वह्नि दीपनं।।[53]

शीत वीर्यं विषालक्ष्मी पापपित्तानिलापहम्।

अल्पाभिष्यन्दि कान्त्योजस्तेजों लावण्य वृद्धि कृत्।[54]

स्वर स्मृति करं मेध्यमायुष्यं बलकृद गुरु।

उदावर्त ज्वरोन्माद शुलानाह व्रणान् हरेत्।।

स्निग्धं कफकरं रक्षः क्षय विसर्प रक्तनुत्।।[55]

घी रसायन, स्वादिष्ट, नेत्रों के लिए हितकर, अग्निदीपक, शीत वीर्य, किंचित् अभिष्यंदि, कान्ति, ओज, तेज और लावण्य की वृद्धि करने वाला, स्वर को स्वच्छ करने वाला, स्मरण शक्ति को वढ़ाने वाला, मेधा (धारणा शक्ति) के लिए हितकर, आयु को वढ़ाने वाला, वल कारक, गुरु, स्निग्ध, कफ कारक एवं विषघ्न है।

गाय के घी के गुण

गव्यं घृतविशेषेण चक्षुष्यं वृष्यमग्नि कृत्।

स्वादु पाक करं शीतं वात पित्त कफापहम्।।[56]

मेधा लावण्य कान्त्यौजस्तेजो वृद्धि करं परम्।

अलक्ष्मी पापरक्षोघ्नं वयसरु स्थापक गुरु।।[57]

बल्यं पवित्रमायुष्यं सुमंगमल्यमं रसायनम्।

सुगन्धं रोचनं चारु, सर्वाज्येषु गुणाधिकम्।।[58]

एक दिन पूर्व के दूध से निकाले हुए घी को संस्कृत में हैयंगवीन कहते हैं। हैयंगवीन नेत्रों के लिए हितकर, अग्नि दीपक, अत्यन्त रूचि कारक, बल वर्द्धक, वृंहण (रस रक्तादिवर्द्धक), वीर्य वर्द्धक है तथा विशेष करके ज्वर नाशक होता है।

घी जिनके लिए निषिद्ध है–

राजयक्ष्मणि वाले च वृद्धे श्लेष्मकृते गदे।।[59]

रोगे सोमे विषूयांच विबन्धे च मदात्यये।

ज्वरे च दहने मन्दे न सर्पिवर्हु मन्यते।।[60]

वालक तथा वृद्ध के लिए, राजयक्ष्मा (टी.वी.), कफ जन्य रोग, आम युक्त रोग, विषूचिका (हैजा), मलवन्ध, मदात्यय, ज्वर तथा अग्नि की मन्दता इन सब स्थितियों में घी प्रशस्त नहीं है।

प्रस्तुत लेख में उद्धृत श्लाकों की सन्दर्भ सूची

क्रम संख्या	श्लोक संख्या	अधान्त शब्द	सन्दर्भग्रन्थ	अध्याय	श्लोक संख्या
1.	1	शरीर्तेन्द्रय.....अयुरुच्यते	चरक संहित (सूत्र स्थान)	01	42
2.	2	दुग्धंसीरं.....सरम्	भव प्रकाश विघुण्य (दुग्ध वर्ग)	दुग्ध वर्ग	01
3.	3	गव्यदुग्धं.....सेविना रुदा	भव प्रकाश विघुण्य (दुग्ध वर्ग)	दुग्ध वर्ग	7
4.	4	जंगलानूप ..... प्रवर्तत	भव प्रकाश विघुण्य (दुग्ध वर्ग)	दुग्ध वर्ग	13
5.	5	स्वल्पान्न.....स्तिम्	भव प्रकाश विघुण्य (दुग्ध वर्ग)	दुग्ध वर्ग	14
6.	6	अल्याभि.....परम स्मृतम्	सुश्रुत संहिता (सूत्र स्थान)	45	47
7.	7	प्रायः प्रभविक.....चापराछिकम्	सुश्रुत संहिता (सूत्र स्थान)	45	59,60
8.	8	पयोभिष्यन्दि.....तद्धितम्	सुश्रुत संहिता (सूत्र स्थान)	45	61,62
9.	9	स्वादु शीतं.....रसायनम्	चरक संहित (सूत्र स्थान)	27	218
10.	10	रोचनं दीपने.....पूंरणंदधि	चरक संहित (सूत्र स्थान)	27	225

क्रम संख्या	श्लोक संख्या	अधान्त शब्द	सन्दर्भग्रन्थ	अध्याय	श्लोक संख्या
11.	11	पीनसे.....स्रोतोकिशोधक	चरक संहित (सूत्र स्थान)	27	226,227,228
12.	12	दधितु.....मंगलय०च	सुश्रुत संहिता (सूत्र स्थान)	45	65
13.	13	महाभिष्यादि.....त्रिदोषकृत	सुश्रुत संहिता (सूत्र स्थान)	45	66
14.	14	स्निग्धंपाके.....रुचिप्रंदम्	सुश्रुत संहिता (सूत्र स्थान)	45	67
15.	15	वातघ्नं.....सुपरिस्रुतम्	सुश्रुत संहिता (सूत्र स्थान)	45	76
16.	16	श्रृतात्.....बलवर्द्धनम्	सुश्रुत संहिता (सूत्र स्थान)	45	77
17.	17	दध्नःसरो.....विषर्द्धनम्	सुश्रुत संहिता (सूत्र स्थान)	45	78
18.	18	दधित्वसारं.....रूचिप्ररम्	सुश्रुत संहिता (सूत्र स्थान)	45	79
19.	19	शरदग्रीष्म.....शस्यते	सुश्रुत संहिता (सूत्र स्थान)	45	80
20.	20	मृष्णाक्लम्.....करोतिच	सुश्रुत संहिता (सूत्र स्थान)	45	81,82
21.	21	दध्युष्णं.....शुक्रकृत्	भाव प्रकाश निघुण्दु	दधि वर्ग	01,02
22.	22	गव्यदधि.....गुणाधिकम्	भाव प्रकाश निघुण्दु	दधि वर्ग	10

क्रम संख्या	श्लोक संख्या	अधान्त शब्द	सन्दर्भग्रन्थ	अध्याय	श्लोक संख्या
23.	23	न वक्तं.....नामलकैर्विना	भाव प्रकाश निघुण्दु	दधि वर्ग	17
24.	24	ज्वरासृक्.....दधिप्रियः	भाव प्रकाश निघुण्दु	दधि वर्ग	19
25.	25	शेफार्शो.....गरेषुच	चरक संहिता (सूत्र स्थान)	27	229
26.	26	मन्थनाद.....घ्रुलमुच्यते	सुश्रुत संहिता (सूत्र स्थान)	45	85
27.	27	रुक्षमर्द्धो.....प्रयोजयेत	तत्राान्तरे	—	—
28.	28	नैव तक्र.....रक्तपैत्तिके	सुश्रुत संहिता (सूत्र स्थान)	45	86
29.	29	शीतकाले.....प्रशस्यते	सुश्रुत संहिता (सूत्र स्थान)	45	87
30.	30	न तक्र सेवी.....तक्रमाहु	भाव प्रकाश निघुण्दु	45	07
31.	31	फोलंतु मथित.....रसालवत्	भाव प्रकाश निघुण्दु	45	01,02
32.	32	घोलंतु मथितं.....पिन्तनुत्	भाव प्रकाश निघुण्दु	45	03
33.	33	तक्रं ग्राहि.....कफायहम्	भाव प्रकाश निघुण्दु	45	04,05,06,
34.	34	उदश्वित्.....लवणान्विता	भाव प्रकाश निघुण्दु	तक्र वर्ग	08

क्रम संख्या	श्लोक संख्या	अघान्त शब्द	सन्दर्भग्रन्थ	अध्याय	श्लोक संख्या
35.	35	वातेम्लं.....युतं कफे	भाव प्रकाश निघुण्दु	तक्र वर्ग	11
36.	36	हिंगुगीर.....सचित्रकम्	भाव प्रकाश निघुण्दु	तक्र वर्ग	12,13
37.	37	तक्रमामं.....प्रयुज्यते	भाव प्रकाश निघुण्दु	तक्र वर्ग	14
38.	38	नैव तक्र.....रक्तपित्तजे	भाव प्रकाश निघुण्दु	तक्र वर्ग	17
39.	39	संग्रहि.....रुचिनाशनम्	चरक संहिता (सूत्र स्थान)	27	230
40.	40	नवनीतप्रनः.....हृद्यं	सुश्रुत संहिता (सूत्र स्थान)	45	92
41.	41	प्रक्षणं सरजं :.....बलग्निकृत्	भाव प्रकाश निघुण्दु	नवनीत वर्ग	01
42.	42	संग्राहिः.....शिशोः	भाव प्रकाश निघुण्दु	नवनीत वर्ग	02
43.	43	दुग्धोत्थम्ः.....शीतलम्	भाव प्रकाश निघुण्दु	नवनीत वर्ग	04
44.	44	दध्नासहः.....तक्र कूर्चिका	निघण्टुरलाकर	—	—
45.	45	गुरुः किलाटोः.....मोरटौ	चरक संहिता (सूत्र स्थान)	45	91
46.	46	संतानिका पनः.....गुर्वीच	सुश्रुत संहिता (सूत्र स्थान)	45	94

क्रम संख्या	श्लोक संख्या	अघान्त शब्द	सन्दर्भग्रन्थ	अध्याय	श्लोक संख्या
47.	47	स्मृति बुद्धि .....ज्वरापहम्	सुश्रुत संहिता (सूत्र स्थान)	27	231
48.	48	सर्व स्नेहोत्तमं.....सहस्रकृत्	चरक संहित (सूत्र स्थान)	27	232
49.	49	मदापस्मार.....व्ययोहति	चरक संहित (सूत्र स्थान)	27	233
50.	50	उग्रगन्धं.....मतःपरम्	चरक संहित (चिकित्सा स्थान)	अध्याय 1	—
51.	51	एकादश.....महाघृतम	सुश्रुत संहिता (सूत्र स्थान)	45	109
52.	52	वियांके मधुरं.....सर्पिगुणोत्तरम	सुश्रुत संहिता (सूत्र स्थान)	45	97
53.	53	घृतमाज्य ..... दीयनम्	भाव प्रकाश निघुण्दु	घृतवर्ग	01
54.	54	शीतवीर्य.....वृद्धिकृत्	भाव प्रकाश निघुण्दु	घृतवर्ग	02
55.	55	स्वर स्मृति कंर.....रक्तनुत्	भाव प्रकाश निघुण्दु	घृतवर्ग	03
56.	56	गव्यं घृतं.....कफायहम्	भाव प्रकाश निघुण्दु	घृतवर्ग	04
57.	57	मेधा लावण्य .....स्थापंकुरु	भाव प्रकाश निघुण्दु	घृतवर्ग	05
58.	58	बल्यं.....गुणाधिकम्	भाव प्रकाश निघुण्दु	घृतवर्ग	06
59.	59	राजयक्ष्मणि..... गदे	भाव प्रकाश निघुण्दु	घृतवर्ग	19
60.	60	रोगे सोमे.....मन्यते	भाव प्रकाश निघुण्दु	घृतवर्ग	20

Chapter 25

Shelf Life Studies on Indian Milk Products

Sudhir Singh

Sr. Scientist,

Indian Institute of Vegetable Research, Varanasi

Introduction

India has emerged as largest milk producer in the world. Half of the milk produced in India is utilized as fluid milk and rest of milk is converted in to traditional indigenous dairy products. Furthermore, only 10-15 per cent of the total milk produced is processed by the organized dairy sector. A large proportion of milk is converted into Indian milk products such as ghee, paneer, chhana and khoa etc.

The indigenous milk products are mainly produced by small scale traders and halwais except for about 15 per cent of total ghee is contributed by the organized dairy sectors. The traditional methods are good for cottage scale which do not permit efficient utilization of fuel. These processes are labour intensive. The quality of Indian milk products varies in respect of moisture, texture and flavour. The quality of milk used, degree of heat treatment, manner of handling of products and duration and condition of storage lead to variations in the chemical composition which pose problems in the production of uniform quality sweets.

The unhygienic conditions at the production centres lead to contamination of products with different types of micro organisms leading to low shelf life of these products. The problem gets aggravated during summer month in respect of composition, microbial quality, sensory properties and shelf life. A variety of micro organisms comprising of mesophiles, acid producers, spore farmers, coliforms, staphylococci, yeast and mould, can be located in indigenous milk products.

Most of the indigenous milk products are sold in the market without proper packaging and unduly exposing to atmospheric contamination. It is thus, not feasible to lay down any microbiological standards for these products.

Shelf Life Studies of Indian Milk Products

Ghee

Ghee has a long shelf life to resist spoilage by elemental and microbial means than any other milk products. However, it is on common knowledge that upon prolonged storage at ambient temperature, it undergoes oxidative changes. Reaction of oxygen with unsaturated fat is a major cause of spoilage. It gives rise to a typical, strong and disagreeable odour. Production of off flavour results in the loss of nutritive value.

Auto oxidation of ghee is aggravated by metallic contamination and sunlight. The acceleration effect of light is dependent on its wave length and visible light accelerates the decomposition of hydroperoxidase. Ultraviolet light has more pronounced effect than the impact of other rays. High energy radiations such as β and γ rays exert a pronounced acceleration effect because the generation of hydroperodixase and free radicals is responsible for off flavour development from the molecules of unoxidized substrate.

The durability of ghee is affected by the degree of unsaturation of the fat, the temperature at which the ghee is stored, the manner in which the milk of ghee making was handled, uncontrolled fermentation during curdling, uneven heating during manufacture and insanitary conditions of the vessels used for the production and storage of ghee.

A number of synthetic antioxidants such as gallates (ethyl, propyl, octyl) butylated hydroxytoluene (BHT), butylated hydroxyanisole (BHA), tertiary butyl hydroquinons (TBHQ), ascorbic acid, α-tocopherol, phospholipids and some natural antioxidants, namely curry leaves, betel leaves, soybean powder, safflower and amla can be added in small amounts with a view to either complete prevention or partial retardation of fat during storage.

The antioxidative properties of curry and betel leaves can be attributed to their phenolic compounds, predominantly hydroxychavicol. The ascorbic acid enriched leaves may act synergistically. Curry leaves and betel leaves also contain many amino acid, which serve as antioxidants. The age old practice of boiling betel and curry leaves with desi butter at the time of clarification helps to improve the flavour, colour and shelf life of ghee.

Khoa

The average shelf life of khoa at 37±1°C, 23-24°C and 5-10°C is reported 7, 10 and 25 days, respectively. In the absence of proper packaging materials, the moisture loss from the product adversely affects the texture and enhances the rate of chemical deterioration such as oxidation and browning. The effect of packaging materials and preservatives have significant effect on the shelf life of khoa. The shelf life of khoa is increased to 13 days at 13°C and to 75 days in the cold storage during storage in the four ply laminated pouches and tin containers. Sterilization of polypacks with gamma radiation using CO^{60} prior to product filling proved to be beneficial. Addition of 0.3 per cent potassium sorbate during the last stage of khoa making increased the shelf life of khoa by another 10 days. Vacuum packing of khoa could enhance the shelf life upto 120 days during refrigerated storage.

The shelf life of khoa can be extended by converting khoa to khoa powder. Khoa is uniformly distributed on aluminium trays and is dried separately in vacuum and atmospheric hot oven at 70°C. The grated khoa is dried in fluidized bed dryer with inlet temperature of 98°C. Khoa can be dried in 30 minutes in fluidized bed dryer. Khoa powder can also be developed by spray drying process. Concentrated milk with 30 per cent total solids is subjected to heat treatments to develop cooked flavour. The heated, concentrated milk is atomized and instantly dried in an Anhydrospray drier with an inlet temperature of 190°C and outlet temperature of 78°C. Khoa powder is cooled to room temperature and is packed in suitable packaging materials.

Khoa powder has a shelf life of 5-6 months at room temperature. Nitrogen gas flushing and addition of BHA further enhanced the shelf life to 11 months.

Paneer

The relatively short life of paneer is a major handicap on the commercial adoption of paneer manufacture. The shelf life of paneer is reported to be only 6 days under refrigerated condition though its freshness is lost within 3 days. At room temperature paneer is not stored for more than one day. The heat treatment to milk during the manufacture of paneer is more than enough to destroy all the pathogenic spoilage micro-organisms. Post manufacturing contamination is responsible for the spoilage of paneer. The spoilage in paneer occurs due to the surface growth of micro organisms. A greenish yellow slime forms on the surface and the discolouration is accompanied by off flavour development during storage. The shelf life of paneer can be further increased by following methods:

Chilling

Rapid chilling of paneer is necessary to arrest the growth of micro organisms. It is necessary for the processor to reduce the paneer temperature to 5-10°C at the time of packaging. Good quality water should be used for chilling of paneer.

Brining

Paneer dipped in 5 per cent brine solution lasts for nearly 20 days as against control paneer is spoiled after 6 days of storage at 8-10°C. The sensory attributes are rated higher for salted samples. Since paneer is mostly salted and spiced before consumption, the salting at the time of dipping can be used in extending the shelf life of paneer.

Use of Chemical Preservatives

A shelf life of 32 days under refrigeration can be achieved when paneer is treated with a combination of delvocid and hydrogen peroxide. Shelf life of 40 days using benzoic acid (1200 ppm) under refrigerated conditions and 20 days at 37°C has been reported. However, shelf life of 36 days at room temperature is achieved by adding sorbic acid to milk (0.15 per cent) and subsequent wrapping of paneer in sorbic acid coated paper.

Freezing

The flavour and appearance of paneer is not affected on frozen of paneer at –13°C and –32°C for 120 days. However, the body and texture deteriorates and the paneer becomes crumbly and fluffy upon thawing. Blast freezing has been used to enhance the shelf life of paneer. The paneer blocks cut into small size are blast frozen at a temperature below –20°C. Paneer can be stored for more than one year without any deterioration in the quality.

Vacuum packaging

Vacuum packaging of paneer in laminated pouches can increase the shelf life of paneer to about 30 days at 6±1°C. The body and texture of paneer is also improved on vacuum packaging as it becomes more compact and shows better sliceability. Paneer packaged in cryovac vacuum packaging and heat treated at 90°C for one minute can maintain the shelf life of paneer of 90 days at refrigeration temperature.

Heat Sterilization of Paneer

The shelf life of paneer is markedly improved at refrigeration temperature. However, the shelf life of paneer is not improved noticeably. Heat sterilization of paneer is an effective treatment for improving the shelf life of paneer at room temperature. Paneer packed in tins along with water/brine and sterilization at 15 psi for 15 minutes can extend the shelf life of paneer to 4 months.

Microwave Drying of Hot Diffused Paneer

Hot diffusion (60°C for 5 minutes in 2.5 per cent sodium chloride and 0.5 per cent potassium sorbate) and cold diffusion (5-10 °C for 6 hours in 2.5 per cent sodium chloride and 0.5 per cent potassium sorbate) process in paneer and subsequent microwave drying (80 per cent power level for 5 minutes) and vacuum packaging is reported to extend the shelf life to 90 days under refrigerated storage. The dried paneer has no fat loss and apparent browning and is rehydrated well at 30°C for 10 minutes.

Hurdle Technology Concept for Preservation of Paneer

The shelf life in paneer is achieved by inactivation of micro organisms. The shelf life extension in paneer is achieved by inherent preservation factors such as extent of heat treatment, water activity, low temperature storage, redox potential, pH etc. These preservative factors act as hurdles because micro organisms would have to jump these hurdles in order to grow and spoil the product. The stability of paneer depends upon the intensity of hurdles present in it. In conventional preservation method, the intensity of one or two of these hurdles is exceptionally increased making it extremely difficult for micro organisms to jump that hurdle. Paneer can be preserved for a long time by heat treatment alone, but there is immense damage to texture and nutritive value. Paneer's keeping quality can also be considerably extended by dehydration process. However the reconstituability of paneer is poor and the texture is adversely affected. However, by using three or more hurdles together, the damage to sensory properties is not only kept minimum but also their synergistic action is exploited for paneer preservation. In hurdle technology, every preservative parameter is kept at optimum level in order to get maximum microbial inactivation by combination of two or more such parameters so that damage to the sensory properties of food is kept to the minimum level.

Shrikhand

Shrikhand, an acid coagulated sweetened milk product is popular in Gujarat, Maharashtra and Karnataka. Shrikhand has fairly long shelf life of 35-40 days at 8°C and 2-3 days at 30°C. The keeping quality depends largely on the initial level of contaminating organisms particularly yeast and moulds. The keeping quality is seriously affected by proteolytic microflora under refrigerated conditions of storage. The spoiled shrikhand samples exhibit off flavours like fruiting, fermented and rancid along with surface discolouration. Addition of 0.5 per cent potassium sorbate prevents the growth of yeast and moulds. Heating of shrikhand at 70°C for 2 minutes further improves the keeping quality of shrikhand.

Conclusion

The need for increasing the shelf life of indigenous dairy products is greatly realized for effective upgrading the technology for production on large scale. Newer innovations have helped in increasing the shelf life of indigenous dairy products considerably. The judicious use of permitted additives is desired for the safety of consumers.

References

1. Rajorhia, G.S. and Sherbon, J.W. (1986). Effect of processing conditions on the melting properties of ghee. Indian Dairyman 38(8), 381.

2. Rajorhia, G.S. (1998). Recent advances in the manufacture of Khoa. Centre of Advanced Studies in Dairy Technology, NDRI, Karnal.

3. Rao, K.J. (1993). Application of handle technology in the development of long life paneer based convenience food. Ph.D. Thesis. NDRI Deemed University, Karnal.

4. Sachdeva, S. and Singh, S. (1990). Shelf life of paneer as affected by antimicrobial agents Part I. Effect on sensory characteristics. Ind. J. Dairy Sci. 43(1): 60.

5. Upadhyay, K.G. and Dave, J.M. (1977). Shrikhand and its technology. Indian Dairyman. 28(9): 487.

Chapter 26

Milk and Milk Production System: Global Scenario

R.K. Pandey
Department of Animal Husbandry and Dairying, BHU, Varanasi, U.P.

Introduction

Milk production and its consumption in the form of fluid milk or other dairy products are important not only in India, but also in other parts of the world for all categories of population, especially vegetarian masses. Milk and dairy products are the main sources of protein and calcium in readily available form. In western countries, milk and dairy products provide up to 75 per cent of total calcium intake. Being easily digestible milk plays an important role in meeting the requirements of many essential nutrients for children, adult scents, elderly people, pregnant and nursing mothers. This is also considered vary useful for those engaged in performing mental jobs. Milk products are also liked by majority of the population. Thus, it is a useful commodity for almost all categories of population.

It may be interesting to know that the post Independence era witnessed a declining trend in milk production because of lack of focus toward Dairy Development. At the same time "model" imperative for development of Indian Dairy Industry was slowly getting evolved in a salient corner of Gujarat, where the small-small producers got organized to earn a remunerative price for their produce through collective ownership. To take forward this "model" National Dairy Development Board got into act in 1965. In 1970, Operation Flood-I, 1979, Operation Flood II and later in 1986 Operation Flood-III were launched and effectively exacted. Today the institutions under the aunbist of these programmes are channelising about Rs 24 Crores of urban money every day to the rural producers.

World milk production has been continuously growing through at a very insignificant annual growth rate of 0.78 during the last two decades (1984-2003) as against the growth rate of 4.07 per cent in India. This general growth in milk production can be observed in all regions of the world except majority of countries is Europe, which registered a negative growth in the past two decades. In particular, the growth in milk production is much more pronounced in case of buffalo, particularly in India.

After successful implementation of operation flood programmes, India is now become the largest milk producing country in the world. This was largely achieved by significant improvement in the productivity of bovines during the past three decades as well as due to increase in bovine population. Currently, India with almost 16.9 per cent of world's human population is contributing nearly 14.7 per cent of milk to the total world milk production of 592 million tonnes in 2003 as against its share of 9 per cent in world milk production in 1970 (Table 26.1).

Present Dairy Scenario

At same time, India also account for largest, nearly 20 per cent of entire world's bovine population. This large scale of operation involves around two activities *i.e.* Dairy Forming and Dairy Processing. Dairy Farming is primarily in the unorganized sector while Dairy Processing is managed by three key players in the milk processing sector. Dominant player is a large base of milk producers Co-operative federations. Private and multinationals are second in terms of milk processing followed by Government owned Dairy Plants.

Milk Production Outlook Across World

With the changing milk production and consumption pattern across different countries of the world, India is likely to emerge as a global player in dairy export provided steps are taken to adhere to phyto-sanitary and quality standards in milk production and its processing.

Table 26.1: Fresh Milk Production (Million Tonnes)

Country	1997	1998	1999	2000	2001	2002
USA	70.80	71.41	73.80	76.06	75.05	75.02
Germany	28.70	28.37	28.33	28.33	28.19	28.10
France	24.91	24.83	24.89	24.89	24.79	25.10
UK	14.84	14.63	15.01	14.48	14.70	14.98
New Zealand	11.05	11.38	10.8	12.23	13.16	13.90
Australia	9.30	9.72	10.48	11.17	10.87	11.62
Netherlands	10.92	10.99	11.17	11.13	10.50	10.45
Ireland	5.25	5.09	5.12	5.15	5.34	5.40
Denmark	4.63	4.66	4.66	4.71	4.53	4.60
China	6.67	6.62	7.17	8.27	8.66	9.06
India	66.30	70.80	74.30	78.00	81.90	84.50

Milk has shown a steady increase in the last few years and estimated at 92 million tonnes in the year 2005 end. Some of the key milk rich states, which contribute significantly to the overall members, are mentioned hereunder in Table 26.2.

Table 26.2: Milk Production Trends in Key States (Million Tonnes)

States	2000-01	2001-02	2002-03
Andhra Pradesh	5.52	5.81	6.58
Gujarat	5.31	5.86	6.08
Haryana	4.84	4.97	5.12
M.P.	4.76	5.28	5.34
Maharastra	5.84	6.09	6.23
Punjab	7.78	7.93	8.17
Rajasthan	7.45	7.76	7.79
Tamil Nadu	4.91	4.98	4.62
Uttar Pradesh	13.86	14.65	15.29
West Bengal	3.47	3.51	3.60

FAO latest estimates forecast that Chinese milk production with 21 million tonnes during current year is 20 per cent more than last year, and three times what it was 5 years ago. Last five year, trend validate that milk production increased by 24 per cent in New Zeland, 17 per cent in India and 12 per cent each in Pakistan, Brazil and Mexico. Production has increased by 5 per cent in US and is unchanged in EU and Russia. It is down by 26 per cent in Argentina and by 6 per cent in Australia (due to drought) Table 26.3.

Table 26.3: World Milk Production (Million Tonnes)-Lastest Trend

Country	*2004 Estimate*	*V/s 1999*
EU 15	122.0	Nc
India	91.5	+17 per cent
US	77.2	+5 per cent
Russia	32.0	Nc
Pakistan	29.1	+12 per cent
Brazil	24.2	+12 per cent
China	21.0	+195 per cent
New Zealand	14.6	+24 per cent
Mexico	10.3	+12 per cent
Australia	9.9	-6 per cent
Argentina	7.8	-26 per cent

Preamble

It may be interesting to know that the post independence era witnessed a declining trend in milk production primarily because of lack of focus towards Dairy Development. At the same time, a "*model*" imperative for development of Indian Dairy Industry was slowly getting evolved in a silent corner of Gujarat, where the small-small producers got organized to earn a remunerative price for their produce through collective ownership. To take forward this '*model*', National Dairy Development Board got into act in 1965. In 1970, Operation Flood-I 1979, Operation Flood-II and later in 1986 Operation Flood-III were launched and effectively executed. Today the institutions under the ambit of these programmes are channelising about Rs. 24 Crores urban money every day to the rural producers.

After successful implementation of Operation Flood Programmes, India has now become the largest milk producing country in the world (Table 26.1). At the same time, India also accounts for largest, nearly 20 per cent of entire world's bovine population. This large scale of operation revolves around two activities *i.e.*, Dairy Farming and Dairy Processing. Dairy Farming is primarily in the unorganized sectors while Dairy Processing is managed by three key players in the milk processing sector. Dominant player is a large base of milk producers co-operative unions federated onto 22 state co-operative federations. Private and multinationals are second in terms of milk processing followed by Government owned dairy plants.

Milk Production

Milk production has been shown a steady increase in the last few years and estimated at 86.6 million tones in the year 2003 end. Some of the key milk rich states, while contribute significantly to the overall numbers, are tabulated in Table 26.2.

FAO latest estimates forecast that Chinese milk production with 21 million tones during current year, is 20 per cent more than last year, and three times what it was 5 years ago. Last 5 years trend validated that milk production increase 24 per cent in New Zealand, 17 per cent in India and 12 per cent each in Pakistan, Brazil and Mexico. Production has increased by 5 per cent in US and is unchanged in EU and Russia. It is down by 26 per cent in Argentina and by 6 per cent in Australia (due to drought) as tabulated in Table 26.3.

Dairy Infrastructure

Post liberalization of the dairy processing industry in the year 1991, brought about a significant upsurge in milk and milk products manufacturing capabilities in the country. The increase in dairy processing units had largely been in the small to medium capacity units.

Overall, Dairy infrastructure has expanded considerably as shown in Table 26.4.

Available data show that 33 per cent of total milk production is retained at the village or farm level while balance is shared by the traditional un-organized sector (51 per cent) comprised of small vendors, sweet makers, very small dairy units and the balance (16 per cent) by organized players. Un-organized sector largely deals with loose and unprocessed, liquid milk, traditional sweets, curd, chhana, paneer and loose ghee. Organized players are co-operative unions/federation and private players, engaged in the processed milk, milk powder of various formulations, infant and malted milk foods, processed dairy products, cheese ice creams, butter and ghee.

Table 26.4: Dairy Infrastructure in India

Institutional and Handling Capacity	*Number/ Capacity of Handling*	*Dairy Plants Operating in India*		
		Dairy Plants	*Number*	*Capacity ('000L/d)*
State Coop. Federation	22	Cooperative	212	28394
District Milk Federation	180	Private	403	32415
Dairy Cooperative	84289			
Rural milk procurement	15780 tonnes/d	Others	63	12170
Milk marketing	9534 tonnes/d	Total	67	72979
Processing capacity (liquid)	27084 tonnes/d			

Export

Dairy export though picking up, have still a long way to go as it stands at 1 per cent of total livestock exports and thus insignificant of value terms (Table 26.5)

Table 26.5: Dairy Exports (Rs. In Million)

Dairy Products	*Value*
Milk and Cream (not concentrated)	0.6
Milk and Cream (concentrated)	736
Butter milk, yoghurt, cream	0.9
Butter, Butter oil	237.4
Whey and whey products	37.1
Cheese and curd	7.5
Total Dairy Products	1019.5

World dairy commodity prices fluctuate quite significantly and at time cyclic in nature (Table 26.6)

Table 26.6: World Dairy Commodity Export Prices

Commodity	*April 2001*	*February 2002*	*March 2002*	*April 2002*
SMP	2038	1574	1473	1371
WMP	2000	1549	1488	1416
Acid Casein	5300	4600	4449	4280
Cheddar Cheese	2025	1987	1924	1880
Butter	1275	1094	1036	1001

Source: FAO Food Outlook May 2002.

Recent six monthly review of world dairy situation by USDA suggest that world prices are probably at the peak and likely to moderate for rest of the year. World supply and demand as per USDA is as shown in (Table 26.7).

Table 26.7: World Top Dairy Companies

Company	*Country*	*Dairy Sales Billion USD July 2003*
Nestle	Swi	15.3
Dean Foods	USA	7.1
Dairy farmers of USA	USA	6.4
Aira Foods	Den/Swe/UK	6.1
Danone	France	6.0
Fonterra	New Zealand	5.8
Kraft Foods	USA	5.3
Lactalis	France	5.2
Unilever	NL/UK	4.9
Bongrain	France	3.7

International Collaborations

Some of the top Dairy companies of the world (Table 26.8) have shown business interest in India in the recent past. Nestle has increased its equity base in India and expanded its products portfolio particularly fresh dairy products. Danone is actively considering entirely India with its Yoghurt range. Fonterra in New Zealand has formed a JV with Britannia. Unilever is in Frozen Desserts though exited from milk powder and Fat business. Bongrain has business tie-up with Dabur primilarly in Cheese and retail products. Schereiber Foods has taken majority stake in Dynamix. Nutricia acquired milk products business of Hindustan Livers but has exited. Candia tied up with Jaipurias for Ice Creams. Baskin Robins entered and intend to established International players like Kraft Foods and Lactalis are evaluating business proposals with domestic players.

Table 26.8: Key Domestic Players

Company	*Dairy Brand*	*Dairy Categories*
GCMMF	Amul	Butter, Cheese, Liquid milk (Fresh and UHT), Ghee, Dairy Powders etc.
Mother Dairy	Mother Dairy	Liquid milk, Curd, Flavoured milk, Butter, Ice Cream, Dairy whiteners and Cream
Smith Klein Beecham	Horlicks Viva, Boost, Maltova	Malted Food Drink
Nestle	Nestle	Dairy Powders, Ghee, UHT milk, Curd, Butter and Infant Foods.
Milkfed	Verka	Liquid milk, Curd, Paneer, Milk Powder
BNZF	Britania Milkman	Cheese, Dairy Whitener, Butter and Ghee.
Ved Ram and Sons	Paras	Liquid milk, Curd, Paneer& Ghee
Hatsun Agro	Arofhya	Liquid milk and Ice Cream
Cadbury's	Bournvita	Malted Food Drink
Heritage Foods	Heritage	Liquid mlk, Ghee and Butter
Uniliver	Kwality Walls	Ice Cream

Domestic Players

Key domestic players having sizeable scale of operations in co-operative as well as private sector are tabulated in Table 26.9. Nevertheless, there are other local co-operative federations with established regional brands like *Saras, Vita, Parag, Nandini, Aavin, Vijaya,* etc. having a broad based dairy products portfolio. Prominent labels are *Heritage, Milkfood, Nova, Mahaan, Vadilal,* etc.

Table 26.9: Dairy Products Categories

Traditional	*Value Added*
Paneer/Chhana	Packaged Fresh Dairy Products
White Butter	Milk Powders
Ghee	Cheese
Khoa	Infant Foods
Cream	Malted Foods
Loose milk/Dahi	UHT milk
Sweets	Pasteurized Milk, Condensed Milk, Table Butter, Sterilized Flavoured Milk.

Table 26.10: Continent-wise Milk Production for Different Species in 2003 (Thousand Tonnes)

Continent	*Milk Production*			
	Cow	*Buffalo*	*Goat*	*Total*
Europe	210565 (41.54)	144 (0.20)	2433 (20.30)	213142 (36.03)
Asia	104685 (20.65)	70463 (96.94)	6404 (53.42)	181552 (30.69)
North America	99108 (19.55)	—	174 (1.45)	99282 (16.79)
South America	46312 (9.14)	—	183 (1.53)	46495 (7.86)
Africa	21118 (4.17)	2077 (2.86)	2793 (23.30)	25988 (4.39)
Oceania	25065 (4.95)	—	—	25065 (4.24)
World	506852	72685	11987	591524

Branded Dairy Products

Branded Dairy Products in consumer packs have evolved over a period of time and cover quite a wide range of products categories *i.e., Cultured, Condensed or Heat desiccated, Acid Precipitated and Frozen* as tabulated in Table 26.11.

Table 26.11: Cheese Production of Top Ten Countries in the Year 2003 (Thousand Tonnes)

Countries	Cheese Production
USA	4224
France	1773
Germany	1765
Italy	1099
Netherlands	642
Poland	520
Russia Fed.	500
Egypt	498
Argentina	450
UK	396

Products Categories

Dairy products categories can be grouped as Traditional and Value Added. Traditional products are mainly in the unorganized sector (organized sector has entered in a limited way) and includes a wide range of products categories

Dairy Market

Dairy Market in India, though not fully established, has evolved over a period of time. In nutshell, it can be stated that:

- ☆ Traditional dairy products have mass appeal and offer a wide range and diversified products.
- ☆ Consumption base is fairly large and familiarity/understanding of milk and milk products is deep.
- ☆ Milk is perceived to be highest from of nutrition and it is a part of every diet/meal in one from or the other.
- ☆ Products categories are price sensitive and thus large volumes are critical.
- ☆ Gradual shift from loose to packaged curd are few examples.
- ☆ Preference stays for India or more especially regional taste.
- ☆ Product category architecture is broad based.
- ☆ Fresh dairy products are emerging as categories of future.

- ☆ Milk short and milk surplus areas are geographical different.
- ☆ Regional preferences and usage of product categories are different (For instance Mumbai is a cheese market North-East is primarily milk powder consumption area).
- ☆ Many categories are just commodities like Pure Ghee and Milk Powders and are prone to violent fluctuations in input prices and market conditions.
- ☆ New products and new innovations are being introduced in the market place. Significant changes are being experienced in the Dairy Industry Sector as well as in the market place.
- ☆ Larger industrial infrastructure of various shapes and sizes has mushroomed post liberalization. Good manufacturing practices yet to be adopted in many dairies and plants.
- ☆ Outsourcing is replacing own manufacturing and thus a clear-cut differentiation is getting established between "Manufactures" and the "Marketers". This was not the trend a few years back.
- ☆ Regulatory framework is undergoing review and the internationally accepted norms are under adoption so far as Food Laws are concerned.
- ☆ Domestic players are giving serious look to exports as international subsidies are easing out.
- ☆ Markets are experiencing a change in its character. Consumes are becoming more demanding on hygiene and food safety.
- ☆ Packaged ethnic dairy products have started finding their way and acceptability in the developed countries. The Canadian IDP Foods, Industries, to manufacture Indian milk products in America is a beginning. "Haldiram" brand and packaged Indian dairy based products presence in Asian and UK market is indication of potential the category possesses. In the domestic market, *K.C.Das, Bikanerwala, Chitales, Bikaji and Ganguram* have successfully demonstrated manufacturing and distribution of such products though in a limited way.

Challenges

The positive aspects that the Indian dairy Industry processes are overshadowed by many challenges, which are really structural issues; like

1. High fragmented back end of the business. Low milk yield of cattle attribute to infrastructural and low technology farm management practices.
2. A limited volume of total production is tapped for processing and value addition. In many developed countries, as high as 98 per cent of milk is processed (Institutional average 52 per cent) whereas only 15-16 per cent of the produce is processed in India.
3. Wastages are high due to a variety of reasons primarily climatic conditions coupled with poor cold chain.
4. Fiscal issues like unfavourable central as well as local taxes.
5. Parallel economy thriving on adulteration of milk by replacing milk solids with cheaper ingredients.
6. Limited processing technology exists to handle mass consumption products (*Paneer, Khoa and Sweets)* on a commercial scale.
7. Dumping of subsidized imported milk products.
8. Poor sanitary standards and poorly defined last leg of the distribution
9. Large reliance on domestic market and the export is just negligible.

Outlook

Today's negatives shall be tomorrow's positives given the immense opportunity Indian Dairy Industry has. It is hard to anticipate that our country will be able to significantly increase its share in world trade/export market was western dairy products. Nevertheless, export of indigenous dairy products (*Burfi, Milk Cake, Lassi etc*) can get established in the international market like Chinese food and Pizzas provided organized players are focused and put in high on its wish list. The future for commodities like Powder and Ghee does not appear sustainable and thus a complete technological intervention to commercialized mass market products meeting and suiting the needs of export as well as domestic consumers. However, manufacturing of commodities is essential to till the void between lean and flush months and to cater to the requirement of milk short area.

Domestic market is tough and, therefore, International collaborations or joint ventures have not found feet firmly on ground so far. However, a process of technology and capital transfer has commenced which may help domestic players to imbibe best proven international practices.

Dairy farming and diary processing have overgrown and created a situation of surpluses. This has posed a difficult challenge for dairy marketing and distribution. The last leg of the supply chain has enormous task ahead to convert this situation of plenty into a situation of advantage. It is easier said than done but has become imperative and critical for the industry to take a leap forward.

Dairy farming policies and Food loss are changing to keep the Industry competitive and fit. The dairy processing got delicenced long back, Milk and Milk products Order has lifted embargo on milk shed areas, Food laws are adapting Codex Alimetarius and Tariff barriers are becoming softer. But moch more is required to be done at a much quicker pace.

Products, Processing and Packaging innovations will certainly drive the market based on right consumer insights and appropriate intended use of dairy products. The industry can ill afford to revolve around two or three basic product categories, which are largely trade oriented and are cyclic in nature. In-house as well as research organizations have to further their connectivity with consumers the research has to become more application oriented.

Given the prevailing scenario, emergence of a healthy and robust Dairy Industry in India does not seem to be a *hundred metre race* but a *marathon.*

Acknowledgement

The information, which is in public domain, has been freely gathered from various sources and acknowledged.

Index

C

D